A GLOBAL TOUR

# HOW THE WORLD GETS HIGH

## Sacred Mushrooms, Amazon Vines, Mad Honey & Desert Toads

*and the New Science of Psychedelic Healing*

TJ TOCE

# COPYRIGHT

**HOW THE WORLD GETS HIGH**

*A Global Tour of Sacred Mushrooms, Amazon Vines, Mad Honey, Desert Toads and the New Science of Psychedelic Healing*



First edition, 2026

Published by Beverly Hills Press, Beverly Hills, California

ISBN 979-8-9977099-0-7 (paperback)

ISBN 979-8-9977099-1-4 (hardcover)

Cover design and interior illustrations by the author.

Interior maps, charts, timelines and diagrams are original works created by the author.

Historical illustrations reproduced in this book are in the public domain. *Bufo alvarius* plate: Mary C. Dickerson, *The Frog Book*, Doubleday, Page & Company, 1906, Color Plate V, figure 4.

## IMPORTANT NOTICE

**This book is a work of history, natural history and science journalism. It is provided for informational and educational purposes only.**

Nothing in this book constitutes medical, psychological, therapeutic or legal advice. The author is not a physician, a therapist, a pharmacologist or an attorney.

Many of the substances described in these pages are controlled under federal law in the United States and under the laws of most other countries. Descriptions of how substances have historically been obtained, prepared or used are historical and ethnographic in nature and are not instructions, endorsements or encouragement to obtain, prepare or use anything.

Several substances described here are dangerous. Some can cause serious injury or death, particularly in combination with prescription medications or in the presence of underlying cardiac or psychiatric conditions. Wild plant and fungus identification carries a genuine risk of fatal misidentification.

Legal status varies widely by country, state and municipality, and changes frequently. Any legal information in this book reflects the author's understanding at the time of publication and should not be relied upon. Verify current law in your own jurisdiction.

Readers are responsible for their own choices. The author and publisher disclaim any liability arising from the use or misuse of the information contained in this book.

**If you are struggling with your mental health or with substance use, please speak to a qualified professional.**

# DEDICATION

*To Michael Pollan, for showing me how to change my mind!*

# EPIGRAPH

**"The mind once enlightened cannot again become dark."**

**THOMAS PAINE** *A Letter Addressed to the Abbé Raynal*, 1782

# CONTENTS

# PREFACE

Many of us think of space as the final frontier. To me, the mind has always been the real frontier that leads to all roads. From a very young age, the exploration of the mind and its mysteries has fueled my desire to live a life filled with many wonderful esoteric adventures.

One great truth: no matter where you go on this blue planet you will find something in nature that is designed to expand your consciousness. That is not by accident.

Let me show you the possibilities...

# A NOTE BEFORE WE BEGIN

A few words about what this book is, and what it is not.

**It is a work of history and natural history.** Ten thousand years of it, spread across every continent people have ever lived on. Most of what follows happened long before anyone thought to make it illegal, and a great deal of it is stranger and more wonderful than the modern conversation about psychedelics would lead you to expect.

**It is not a manual.** You will read about how a Mazatec healer prepares mushrooms, how ayahuasca is boiled over three days in the Ucayali, how a Gurung honey hunter climbs down a bamboo ladder two hundred feet above a Nepali valley. Those are descriptions of what humans do and have done. They are not instructions, and they are not an invitation.

**It is enthusiastic, and it tries very hard to be honest.** Those two things pull against each other more often than you might think, and where they do, honesty wins. When a beautiful story is probably a myth, this book tells you the story and then tells you it is probably a myth. When something is dangerous, it says so plainly rather than burying it. When the evidence is thin, it says the evidence is thin.

It is the whole editorial standard, and it comes from a saying I have always liked: you can build your house on sand or on rock, and both look fine on a clear day. You find out which one you built when the rain comes.

**A word about the people this knowledge came from.** Nearly everything in these pages was discovered, preserved and protected by indigenous communities, very often through centuries of persecution for doing so. A great deal of it was then taken without being asked for and without anyone being paid. That story runs through the whole book, and it appears in the first chapter for a reason.

**And a word about safety.** Several of the substances described here can seriously hurt you. Some can kill you. Wild fungus identification in particular is the single most genuinely dangerous activity in this book, and it has nothing to do with the psychedelic and everything to do with the lookalike growing beside it. Where risk is real, you will find a box marked THE STRAIGHT DOPE that gives it to you straight.

Facts first. And then, I promise, the wonder.

**TJ Toce** Beverly Hills, 2026

# HOW THE WORLD GETS HIGH

HOW THIS BOOK TRAVELS

# Six Kingdoms, One Door

*Not country by country, but kingdom by kingdom*

**FUNGI**
psilocybin
*~200 species*

**PLANTS**
DMT
mescaline
*1,000+ species*

**AMPHIBIANS**
5-MeO-DMT
*one desert toad*

**FISH**
from algae
*entirely by accident*

**BEES**
grayanotoxin
*by way of a flower*

**HUMAN BODY**
endogenous DMT
*already installed*

**THE HUMAN NERVOUS SYSTEM**
principal door: the 5-HT2A receptor
*which is yours, not theirs*

*Unrelated branches of life, separated by hundreds of millions of years, arriving independently at molecules shaped closely enough to fit a lock that was already waiting inside your head.*

**THREE HONEST EXCEPTIONS**

Amanita muscaria works on GABA-A. Salvia divinorum works on the kappa-opioid receptor. Grayanotoxin works on sodium channels.

*Please refer to Chapters Two, Three and Six for greater details.*

# CHAPTER ONE

## THE OLDEST APPETITE

*Psychedelic use throughout history, all over the world*

> *Our normal waking consciousness, rational consciousness as we call it, is but one special type of consciousness, whilst all about it, parted from it by the filmiest of screens, there lie potential forms of consciousness entirely different.*
>
> **WILLIAM JAMES The Varieties of Religious Experience, 1902**

There is a plateau in southeastern Algeria called Tassili n'Ajjer, and it is one of the emptiest places on Earth. Sandstone spires. Wind. Rock that holds the day's heat long after the sun is gone. Nothing much lives there now. Dunes, desert and relentless heat.

But it was not always like this. Eight or nine thousand years ago the Sahara was green and lush. There were lakes with hippos in them. There was grass, and there were cattle, and there were people who painted what they saw on the sheltered rock faces of that plateau, and some of those paintings are still there.

One of them shows a man running.

He is holding something in each hand. From his body, and along the outline of his arms and legs, sprout dozens of small, rounded shapes on slender stalks. They look, unmistakably, like mushrooms. They are drawn with care. Whoever made this figure wanted you to know exactly what those things were.

Nobody can tell you for certain what that painting means. Archaeologists argue about it, and they should. Rock art is notoriously slippery, and the history of interpreting it is littered with confident people who turned out to be projecting. Maybe those are mushrooms. Maybe they are something else entirely and we are seeing what we want to see, the way you see faces in clouds.

But here is the thing that makes it hard to dismiss. Consider this. Four thousand miles northwest, in a valley in eastern Spain called Villar del Humo, there is a rock shelter with a painting on it that is roughly six thousand years old. It shows a bull. And beneath the bull, in a neat row, thirteen small, stalked objects with domed caps. In 2011 a team of researchers published a paper arguing that the closest botanical match for those shapes is *Psilocybe hispanica*, a small psychoactive mushroom that grows in the mountain pastures of that exact region, in cattle dung, to this day.

Two paintings. Two continents. Thousands of years and an entire sea between them. No shared language, no trade route, no possibility whatsoever that anyone compared notes.

And that is the pattern this whole book is about.

## THE THING NOBODY TAUGHT US

Humans have been deliberately altering their consciousness since the dawn of time. Not occasionally. Not as a deviation. Constantly, everywhere, in every climate, on every landmass we have ever occupied.

This is not the story most of us grew up with. Most of us absorbed a version where psychedelics arrived in about 1965, wearing a paisley shirt, and left again around 1972. A brief and slightly embarrassing detour in Western history. A thing your uncle did.

That story is wrong in a way that is almost hard to overstate. The 1960s did not invent this. The 1960s interrupted it.

What actually happened is that a practice older than agriculture, older than writing, older than the wheel, and very likely older than religion, ran continuously through nearly every human culture on Earth for thousands of years. Then, in a single decade in the middle of the twentieth century, a handful of governments made most of it illegal, and a great deal of accumulated human knowledge went underground or was simply lost. Sadly, but as it turns out, not permanently.

We are in fact, right now, in the middle of digging it back up.

CHAPTER ONE

# Ten Thousand Years of Getting High

*Documented sites of ancient and traditional psychoactive use*

1. **Tassili n'Ajjer, Algeria** *rock art, c. 7000 BCE*
2. **Selva Pascuala, Spain** *rock art, c. 4000 BCE*
3. **Shumla Caves, Texas** *peyote, 3700 BCE*
4. **Eleusis, Greece** *the Mysteries, 1600 BCE*
5. **Indus Valley** *soma, Rig Veda, 1500 BCE*
6. **Chavín de Huántar, Peru** *San Pedro, 1200 BCE*
7. **Pontic Alps, Turkey** *mad honey, 401 BCE*
8. **Cueva del Chileno, Bolivia** *ayahuasca bundle, c. 1000 CE*
9. **Hawai'i** *'awa arrives by canoe, c. 400 CE*
10. **Kamchatka, Siberia** *fly agaric*
11. **Huautla de Jiménez, Mexico** *teonanácatl*
12. **Ucayali, Peru** *ayahuasca*
13. **Gabon** *iboga, Bwiti*
14. **Annapurna, Nepal** *cliff honey*

How the World Gets High · TJ Toce

---

## SOMA, AND THE OLDEST UNSOLVED MYSTERY IN BOTANY

Around 1500 BCE, somewhere in the northwest of the Indian subcontinent, people began composing the hymns that became the Rig Veda. It is one of the oldest surviving texts in any Indo-European language, and it runs to more than a thousand hymns.

An entire book of it, more than a hundred hymns, is devoted to a single substance called soma.

Soma is pressed from a plant. It is filtered through wool. It is mixed with milk. It is drunk by priests and it is offered to the gods, and the hymns describe its effects in language that is not subtle:

*We have drunk the soma, we have become immortal, we have gone to the light, we have found the gods.*

Elsewhere the poets describe feeling enormous, feeling as though they could lift the earth, seeing the two worlds at once. This is not a beverage. This is a report.

And sadly nobody for certain knows what it was.

The plant is gone, or the identification is lost, or the name migrated to something else. For two centuries scholars have been arguing about it, and the candidate list is long and genuinely interesting. Some argue for ephedra, a stimulant shrub. Some argue for Syrian rue. Some argue for a species of milkweed. And in 1968, an amateur mycologist and retired banker named R. Gordon Wasson published a book making the case that soma was *Amanita muscaria*, the red-and-white fly agaric mushroom of the northern forests.

Wasson's argument is contested and probably not right. But hold on to his name, because he is going to walk back into this story in a way that changes everything.

The point of soma is not the answer. The important point is the question. Three and a half thousand years ago, the founding religious literature of a civilization of tens of millions of people was organized around a plant that reliably produced a delicious and extraordinary state of mind, and the people who drank it considered it so central to their spiritual life that they made it a god. All hail Soma!

---

## THE SECRET THAT HELD FOR TWO THOUSAND YEARS

Through time and space we now arrive fourteen miles northwest of Athens, at a place called Eleusis. There was a sanctuary there where, once a year, for something on the order of two thousand years, several thousand people took part in a ceremony.

We know who went. Plato went. So did Aeschylus and Sophocles and Pindar. Cicero went and wrote afterward that Athens had produced nothing greater. The emperor Marcus Aurelius went. Roman senators went.

Slaves and women and foreigners could go, which in the ancient Mediterranean was remarkable in itself.

We know they fasted for days first. We know they walked the fourteen miles from Athens in procession. We know that at the climax of the rite they drank a barley preparation called the kykeon, and that afterward they entered a hall and something happened there, and that they came out changed. A change for the better no doubt.

And that is where our knowledge stops. Because the one rule of Eleusis was that you did not talk about Eleusis, and for two thousand years essentially nobody did. The penalty was death, but that is not really why it held. It held because the people who went there did not want to cheapen it. Sacred, special and deeply personal.

What we have instead are the things they said around the edges. Cicero: that the Mysteries taught him "not only to live happily, but to die with a better hope." Pindar: that whoever has seen it "knows the end of life and its beginning." Over and over, across centuries, the same note. Not "I had a good time." Something closer to *I am no longer afraid.*

In 1978, Wasson (wow, there he is again) teamed up with a Swiss chemist and a classicist to publish a book arguing that the kykeon was made with barley infected by ergot, a fungus that grows on cereal grain and produces alkaloids closely related to LSD. It is a genuinely elegant theory. It is also unproven and likely to stay that way, because the sanctuary was destroyed in 392 CE when the Christian Emperor Theodosius shut it down, and nobody left us a recipe.

But whatever was in that cup, several thousand of the most literate, skeptical, articulate people of the ancient world went out there year after year for twenty centuries, and every single one of them came back saying the same thing.

That is not nothing.

---

## FLESH OF THE GODS

When the Spanish arrived in Mexico in the sixteenth century, they found something that alarmed them considerably more than gold.

The Nahua peoples had a word: *teonanácatl*. It comes from *teotl*, meaning god or sacred, and *nanácatl*, meaning mushroom. The friars translated it as "flesh of the gods," and the phrase horrified them, because they recognized the shape of it immediately. Here was a people who ate a sacred substance in a ceremony, and were thereby brought into direct contact with the divine, without a priest, without a church, and without any need whatsoever for Rome.

Bernardino de Sahagún, a Franciscan who spent decades documenting Nahua life and left us the extraordinary Florentine Codex, wrote it down carefully even as he condemned it. He described the mushrooms being eaten with honey at feasts. He described people weeping, dancing, seeing visions, and afterward sitting together and telling each other what they had seen.

The Church banned it. The Inquisition prosecuted it. Colonial authorities punished it. And it went into the hills, where it stayed, quietly, for four hundred years.

Not gone. Just waiting.

---

## THE REINDEER AND THE RED MUSHROOM

Go about as far from central Mexico as it is possible to get and you arrive in Siberia, among the Koryak and the Chukchi and the Kamchadal, in a landscape of birch and larch and a great deal of snow.

Here the sacrament is *Amanita muscaria*, the fly agaric, which is the magic mushroom in every fairy tale illustration ever drawn: brilliant scarlet cap, white flecks, thick white stem. It grows in association with birch and pine across the entire northern hemisphere. It does not contain psilocybin. Its active compounds are muscimol and ibotenic acid, they work on an entirely different receptor system, and the experience they produce is genuinely different from a mushroom trip, more dreamlike and more physically peculiar.

Russian explorers in the 1700s came back with accounts that nobody quite believed. The shaman eats the mushroom. The shaman goes into the visionary state. And then the shaman's companions drink the shaman's urine, because the active compound passes through the body largely intact and the toxic components do not, which means the second cup is safer than the first. This was, in a poor and cold place where the mushroom was scarce, a completely rational allocation of a limited resource.

There is also this. Reindeer eat fly agaric. They actively seek it out and they behave strangely afterward. Reindeer herders knew this. And there is a much-loved theory that a red-and-white sacrament, gathered under evergreen trees, delivered by a man in a fur-trimmed coat who comes down through the roof because in a snowed-in Siberian yurt the smoke hole IS the door, and who is towed across the winter sky by intoxicated flying reindeer, might have something to do with a certain December visitor.

Is that true? Honestly, probably not, at least not in the tidy way it gets told. Folklorists are skeptical and they have good reasons. But it is far and away the best story in this chapter, and the fact that it is even arguable tells you how deep these roots run.

---

## THE VINE AND THE LEAF

In 2019, researchers published an analysis of a ritual bundle recovered from a rock shelter called Cueva del Chileno, in the Bolivian Andes, at around thirteen thousand feet. The bundle was made from three fox snouts sewn together. Inside were carved wooden tubes, a headband, and traces of what had been carried in it.

They ran the residue. It came back positive for DMT, harmine, bufotenine, and cocaine.

The bundle is roughly a thousand years old. And harmine plus DMT is the exact chemical signature of ayahuasca.

Now consider what it took to arrive at that recipe. Ayahuasca is made from two plants. One is a woody vine, *Banisteriopsis caapi*. The other is a shrub, usually *Psychotria viridis*, whose leaves contain DMT. On its own, the DMT does nothing when you swallow it, because an enzyme in your gut destroys it before it can reach your bloodstream. However, this amazing vine contains compounds that switch that enzyme off. Combine them, and sacred alchemy begins and the leaf becomes active. Use either one alone, and you have a bitter drink and a wasted afternoon.

The Amazon basin contains something in the neighborhood of eighty thousand plant species.

Ask an ethnobotanist how this was discovered and you will get careful language about long empirical traditions and generations of accumulated trial. It is a reasonable answer and it is probably part of the truth.

Ask a shaman on the Ucayali and you will get a simpler one. The plants told us.

You can decide for yourself which explanation is stranger.

## THE CACTUS THAT WALKED NORTH

In the Chihuahuan Desert, along the Rio Grande, there is a small blue-green cactus with no spines that grows so slowly it can take a decade to add half an inch. It is called peyote, *Lophophora williamsii*, and it contains mescaline.

Dried peyote buttons recovered from the Shumla Caves in south Texas have been radiocarbon dated to roughly fifty-seven hundred years ago. That is older than the pyramids at Giza. That is older than writing.

Peyote use spread north out of Mexico across the plains in the eighteenth and nineteenth centuries, arriving among the Comanche, the Kiowa, the Apache and eventually far up into the northern plains, at exactly the period when those nations were being systematically destroyed. It became the center of a new religious movement that combined indigenous practice with elements of Christianity, and in 1918, in Oklahoma, that movement formally incorporated as the Native American Church.

Which means something worth pausing on. In a country where every classical psychedelic is a Schedule I controlled substance, there is one that several hundred thousand people can legally consume in ceremony, protected by federal statute, and have been able to for decades.

We will come back to peyote properly in Chapter Three, including the fact that it is now in serious conservation trouble. For now, just hold the number. Fifty-seven hundred years, woven through time and space.

CHAPTER ONE

# The Long Continuity

*Ten thousand years of human psychedelic practice, in three acts*

**THE DEEP PAST**

| | |
|---|---|
| 9000 BCE | Tassili n'Ajjer rock art |
| 4000 BCE | Selva Pascuala |
| 3700 BCE | Shumla Caves peyote |
| 1600 BCE | Rig Veda composed |
| 1500 BCE | Eleusinian Mysteries begin |

**THE CLASSICAL WORLD**

| | |
|---|---|
| 1200 BCE | Chavín de Huántar |
| 401 BCE | Xenophon's soldiers |
| 65 BCE | Mad honey ambushes Pompey |
| 392 CE | Eleusis closed by Theodosius |
| c. 400 CE | 'Awa reaches Hawai'i |
| 1000 CE | Cueva del Chileno bundle |
| 1550 CE | Sahagún records teonanácatl |

**THE MODERN ERA**

| | |
|---|---|
| 1918 | Native American Church |
| 1938 | LSD synthesized |
| 1943 | Bicycle Day |
| 1955 | Wasson at Huautla |
| 1970 | Controlled Substances Act |
| 2000 | Johns Hopkins restarts |
| 2020 | Oregon Measure 109 |
| 2026 | FDA priority review |

*Each band has its own scale. Fifty-six years of prohibition, set against ten thousand years of practice.*

How the World Gets High · TJ Toce

---

## 1955: THE YEAR THE DOOR OPENED THE WRONG WAY

In the summer of 1955, R. Gordon Wasson, a vice president at J.P. Morgan who had spent twenty years obsessively studying the role of mushrooms in human culture, traveled to a town called Huautla de Jiménez, high in the mountains of Oaxaca. Land of mole and other delicacies.

There he met a Mazatec woman named María Sabina.

She was a *curandera*, a healer, and she conducted what her people called a *velada*, a night vigil, in which mushrooms were eaten and sung over and used to diagnose and to heal. She had been doing this since she was a child. Her mother and her grandmother had done it. It was not a performance and it was not a novelty. It was medicine and it was church, and it had survived four hundred years of suppression by going quiet and staying local.

She agreed to let Wasson participate. He later wrote that he had asked her permission and that she had granted it on the understanding that he came in good faith and with a serious purpose. He sat through the night. He ate the mushrooms. He heard her sing.

And in May 1957, he published a fifteen-page photo essay about it in *Life* magazine, under the headline "Seeking the Magic Mushroom."

*Life* had a circulation in the millions.

---

What happened next is the part that gets skipped, and it should not be.

They came. Young Americans came by the busload, and then by the thousands, over the next fifteen years. They came to Huautla and they camped and they took mushrooms and they had no idea what they were participating in and no interest in finding out. Bob Dylan was rumored to have come. The Mexican army eventually came too, to clear the town out.

María Sabina's community turned on her. They believed, with some justification, that she had given away the thing that was theirs. Her house was burned. Her son was murdered. She was jailed briefly. She lost her standing and she spent the rest of her life poor.

Near the end she said something that ought to be carved above the door of every retreat center and clinic operating today:

*From the moment the foreigners arrived, the holy children lost their purity. They lost their force. The foreigners spoiled them. From now on they will no longer work.*

She died in 1985.

This book is a celebration. It is going to spend two hundred pages arguing that this is one of the most fascinating and hopeful subjects in the world, and it means it. But you cannot tell this story honestly without telling that part, because the entire modern Western psychedelic movement, every clinic and every conference and every venture-funded startup, traces back through a single night in a small house in Oaxaca and a woman who was not asked what it would cost her.

CHAPTER ONE

## María Sabina

*Huautla de Jiménez, Oaxaca · 1894 to 1985*

The Mazatec curandera who conducted the velada R. Gordon Wasson attended in 1955, *and who paid for it for the rest of her life.*

*Photograph by Juan Carlos Rangel. Public domain.*

How the World Gets High · TJ Toce

---

### BASEL, AND A BICYCLE

Meanwhile, in Switzerland, a chemist was having a completely different kind of afternoon.

Albert Hofmann worked for Sandoz Laboratories in Basel, and in 1938 he was systematically synthesizing derivatives of ergot alkaloids, looking for a circulatory stimulant. The twenty-fifth compound in the series did nothing interesting in animal testing. He set it aside.

Five years later, in April 1943, he went back to it. He said afterward that he simply had a feeling about it, which is not how chemistry is supposed to work and is exactly how a great deal of chemistry actually works.

While resynthesizing it he absorbed a trace amount through his fingertips and went home early, describing "an extremely stimulated imagination" and "an uninterrupted stream of fantastic pictures."

Being a rigorous man, he decided to test it properly. On April 19 he deliberately took what he calculated to be a very cautious dose: 250 micrograms. A microgram is one millionth of a gram. It was, in fact, roughly double a strong recreational dose, because nothing in human experience had prepared him for how potent this compound was.

Twenty minutes later he asked his laboratory assistant to help him get home. Wartime Switzerland had no cars available. So they rode bicycles.

That ride is now commemorated every April 19 as Bicycle Day, and it is the only holiday in the world founded on a chemist's commute.

Sandoz put the compound on the market in 1947 under the name Delysid and sent it to psychiatric researchers around the world, essentially for free, with a note asking them to report what they found. Over the next fifteen years more than a thousand scientific papers were published on it. It was studied for alcoholism, for depression, for anxiety in the dying, for the treatment of what was then called neurosis. Cary Grant took it in therapy roughly a hundred times and gave magazine interviews saying it had made him a happier man.

For about twenty years, this was ordinary science. Boring, even. Published in journals. Funded by governments.

## 1970: THE DOOR SLAMS

You know roughly what happened next, and the full story is Chapter Eight's job. The compound escaped the laboratory. It became a symbol in a culture war it was never equipped to survive. The CIA got involved in ways that were genuinely appalling. A Harvard psychologist made himself extremely famous and made the research community's position extremely difficult.

And on October 27, 1970, President Nixon signed the Controlled Substances Act. LSD, psilocybin, mescaline, DMT and peyote were placed in Schedule I, the most restrictive category, defined as having a high potential for abuse and NO accepted medical use.

Whatever you think about the politics, look at what that designation did in practice. It did not simply ban recreational use. It made research itself nearly impossible. A university lab needed a special license, a locked safe, and years of paperwork to study a compound that a thousand papers had already been published on.

Serious clinical research in the United States essentially stopped for twenty-five years.

Twenty-five years. Let's throw in a little context for clarity. Think about what else happened in medicine in that window. We sequenced the human genome. We built MRI machines that can watch a brain think. And an entire branch of psychiatry sat frozen in a drawer.

---

## WHAT THIS BOOK IS

Here is the strange and wondrous thing, and it is the very reason this book is organized the way it is.

If you look at the list of what humans have used to alter consciousness, it does not sort neatly by continent or by culture. It sorts by KINGDOM. Fungi found this trick. Plants found it, over and over, on separate branches of the family tree that have not shared an ancestor in a hundred million years. An amphibian in the Sonoran Desert found it. A fish in the Mediterranean found it, more or less by accident. Bees found it, by way of a flower. And your own brain, sitting there right now doing whatever it is doing, manufactures at least one of these compounds on its own. Can you see the pattern? Can you see the intention of creation?

Completely unrelated branches of life, arriving independently at molecules shaped closely enough to fit a receptor in the human head that a mushroom in Oregon and a cactus in Peru and a toad in Arizona can each, in their own way, open the same door.

That is either the most interesting coincidence in biology or it is not a coincidence at all.

So that is how we are going to travel. Not country by country, but kingdom by kingdom. Fungi first, because that is where the deepest human history lies. Then plants, which is the biggest room in the house. Then the animals, the amphibians and the fish, because they are strange and wonderful and almost nobody knows about them. Then honey, which is a plant's chemistry laundered through a bee. Then the human body itself, which turns out to have been holding out on us. Then the laboratory. And then, finally, right now, this year, and what happens next.

Along the way, every subject gets the same four questions answered plainly, in a box you can find at a glance: where on Earth it grows, where it is legal, how long we have been using it, and how people actually get it.

One promise about tone. This book is enthusiastic. It thinks this subject is a genuine wonder and it is not going to pretend otherwise. But enthusiasm without honesty is just marketing, so when something is dangerous it will say so, when the evidence is thin it will say so, and when a beautiful story is probably a myth it will tell you the story AND tell you it is probably a myth. You can build a house on sand or on rock. Both look fine on a clear day. You find out which one you built when the rain comes. Get it?

Ten thousand years ago somebody stood on a rock face in a green Sahara and painted a running man with mushrooms coming out of his arms. Why?

We are still trying to work out what he was trying to tell us. But we are finally, after a very long detour, back to asking. Enlightenment is on the path ahead.

Let's go find out.

# CHAPTER TWO

## FUNGI

*The first kingdom! Mycelium Galactica!*

*The humblest fungus displays a life akin to our own.*

**HENRY DAVID THOREAU Journal, 1858**

---

CHAPTER TWO

## The Blue Bruise

*Psilocybe cubensis, Alachua County, Florida*

A single Psilocybe cubensis growing out of cow manure in a pasture.
*Wherever humans took cattle, this mushroom followed.*

*Photograph by Alexander Abair, CC BY 4.0*

How the World Gets High · TJ Toce

---

Pick one up and it will accuse you.

Press your thumb against the stem of a *Psilocybe* mushroom, set it down, and come back in ten minutes. Where your thumb was there is now a bruise, and the bruise is not brown or grey the way a bruise on an ordinary mushroom would be. It is BLUE. Deep, saturated, unmistakable indigo, like your favorite pair of Levi's® jeans.

Foragers have used this tell-tale sign for as long as anyone has been foraging. It is not a perfect test and we will get to why in a moment, but it is a beautiful one, and for a very long time nobody had the faintest idea what caused it.

Here is what we now know, and it is wonderful. The mushroom carries its active compound in a stable, locked form called psilocybin. Damage the tissue, and enzymes inside the mushroom go to work, stripping a phosphate group off the molecule and converting it to psilocin. Psilocin, like indigo, is unstable in open air. It oxidizes, links up with its neighbors into long chains, and those chains happen to absorb light in exactly the way that produces the color blue.

Just fascinating. The bruise is the mushroom's medicine, reacting to injury, in real time, in a color you can see.

The kingdom Fungi performed this trick first, and it got there long before we did. Welcome to the oldest room in the house.

---

## THE GENUS

There are somewhere in the neighborhood of two hundred known species of psilocybin-producing mushroom. They live on every continent except Antarctica. They are not rare, they are not exotic, and a great many of them are growing within a few miles of wherever you happen to be reading this.

Four of them matter most to our story. Let's explore.

**1. *Psilocybe cubensis.*** The famous one. Pantropical, meaning it follows warmth around the globe, and it is found across the Gulf Coast of the United States, throughout Mexico and Central America, in Colombia and Brazil, in Thailand and Cambodia and Vietnam, in India, and across northern Australia. It has one strong preference in life: it grows in the dung of grazing animals. Wherever humans took cattle, *cubensis* followed, which means this mushroom has been quietly riding along with human agriculture for thousands of years.

2. ***Psilocybe semilanceata.*** The liberty cap. Small, pointed, tan, and by most counts the single most widely distributed psilocybin mushroom on Earth. It grows in temperate grassland across Europe, the British Isles, the Pacific Northwest, Chile, New Zealand and Tasmania. Unlike *cubensis*, it does not grow in dung. It grows in the grass NEAR where animals have grazed, feeding on decaying root material. A field in Wales in October can be full of them.

3. ***Psilocybe azurescens.*** The heavyweight. Found along a narrow stretch of the Oregon and Washington coast, in dune grass and wood debris, and it carries the highest psilocybin content of any species yet measured. It was formally described in 1996. Nature saved something very special for the edge of the continent.

4. ***Psilocybe mexicana.*** The historically important one. This is the little brown mushroom of the Oaxacan highlands, the one used in Mazatec veladas, and the one that Albert Hofmann grew in a laboratory in Basel in order to isolate psilocybin in 1958 and synthesize it in 1959. Two years after Wasson's *Life* article, the sacred mushroom of Mexico had been reduced to a white crystalline powder with a formula. Progress is a complicated word.

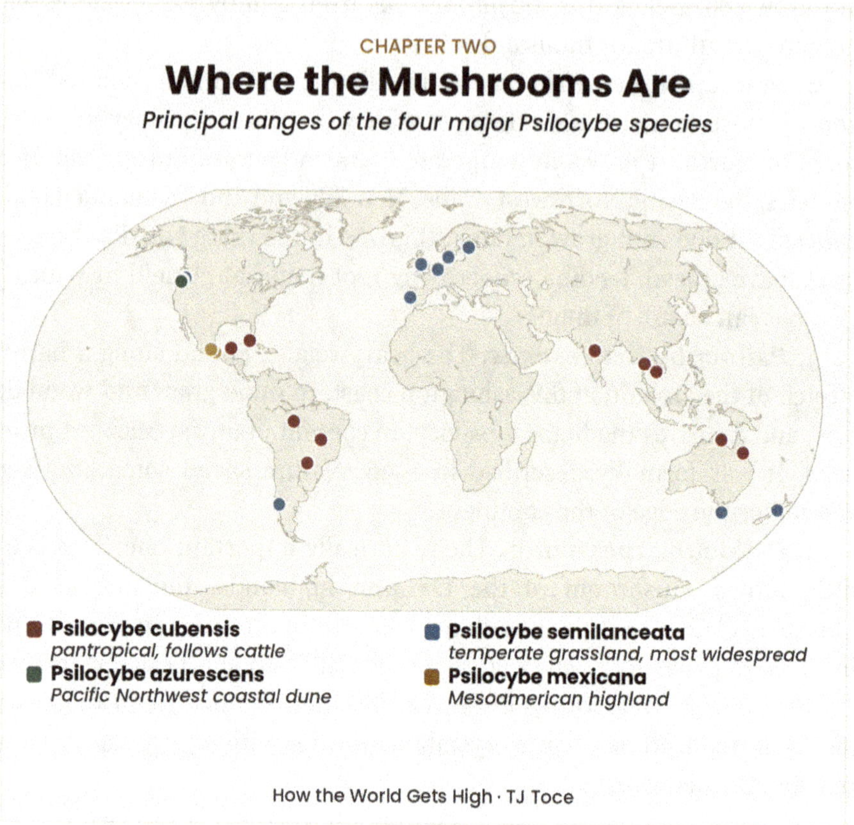


---

## THE CHEMISTRY, MADE SIMPLE

You do not need a chemistry degree for this. You need one picture.

Your brain runs on a set of chemical messengers, and one of the most important is serotonin. Serotonin is involved in mood, in sleep, in appetite, in perception, in how tightly your brain holds on to its own habits. It docks into receptors the way a key fits a lock, and one particular lock, called the 5-HT2A receptor, sits in enormous numbers across the parts of your cortex that handle high-level perception and sense of self.

Now look at psilocin next to serotonin. They are almost the same shape. Same core ring structure, same nitrogen in the same place, a couple of small differences hanging off the edges.

Close enough to fit the lock. Yet, different enough to turn it a wildly different way.

That is the entire secret. When psilocin occupies the 5-HT2A receptor, the parts of your brain that normally run the show in an orderly, predictable, top-down fashion loosen their grip. Regions that rarely talk to each other start talking. The network that maintains your continuous sense of being a separate self, which neuroscientists call the default mode network, goes quiet.

Researchers describe the result as increased entropy in the brain. Humans who have experienced it describe it as the world becoming new again. A grand awakening.

Believe it or not, both descriptions are accurate. Prime Creator at its best.

CHAPTER TWO

# One Shape, Three Molecules

*Why a mushroom fits a lock in your head*

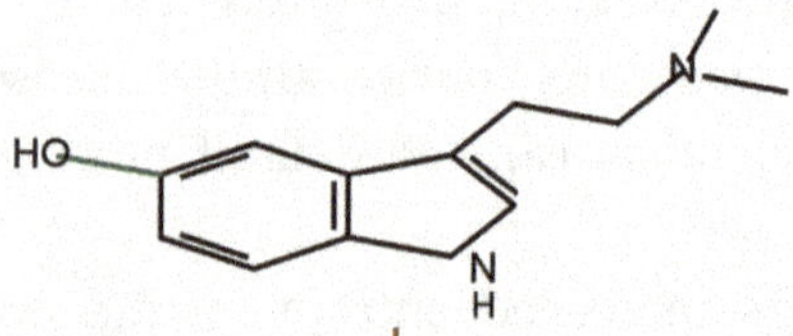


**SEROTONIN**
*5-hydroxytryptamine*
*the one your own body makes*

↓ *eaten*

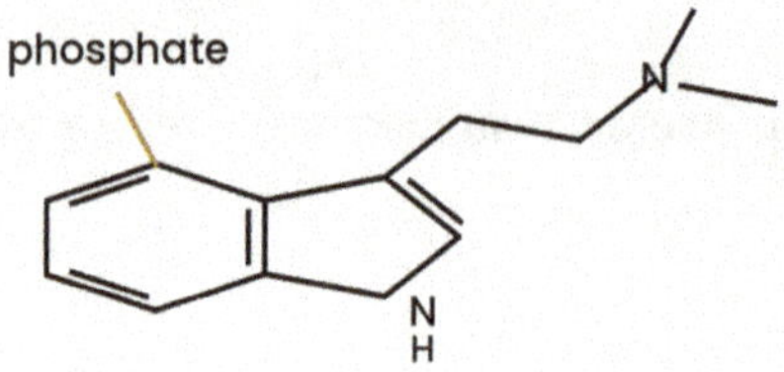


**PSILOCYBIN**
*the stable form inside the mushroom*
*inactive until your body converts it*

↓ *an enzyme strips the phosphate*

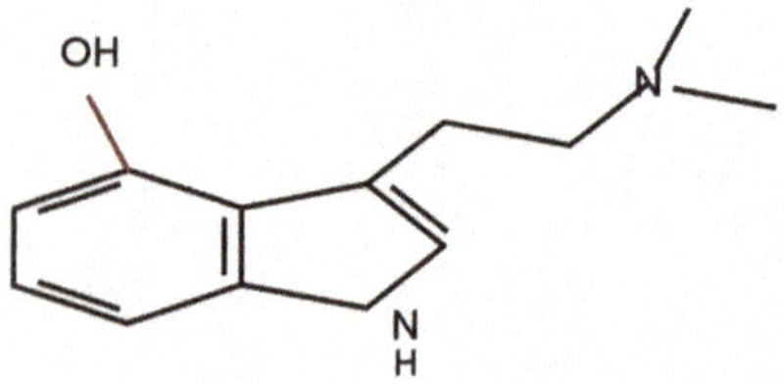


**PSILOCIN**
*what actually reaches the receptor and what turns a bruised mushroom blue*

Look at the shared skeleton. Close enough to fit the serotonin lock. Different enough to turn it a different way.

*That is the entire secret of this book, in three drawings.*

How the World Gets High · TJ Toce

---

## BUT WHY WOULD A MUSHROOM DO THIS?

Consider this, and it is one of the best unanswered questions in all of biology.

A mushroom has no brain. It has no eyes, no nervous system, no opinion about anything. It has no possible use for a molecule that alters consciousness, because it has no consciousness to alter. Manufacturing psilocybin costs the organism real energy. Evolution does not generally fund hobbies.

So what is it FOR?

The leading hypothesis is insect defense. *Psilocybe* species live in dung and decaying wood, which are crowded, competitive neighborhoods full of fungus-eating flies and beetles. A compound that scrambles an insect's nervous system and puts it off its lunch would be worth every calorie. There is decent support for this. Psilocybin appears to suppress feeding behavior in some insects. I'm not convinced.

The second hypothesis is stranger and more delightful. Genetic analysis suggests that the cluster of genes responsible for making psilocybin has jumped between unrelated fungal species, moving sideways rather than being inherited down the family tree. This is called horizontal gene transfer, and it happens most readily where different species live crammed together in the same dung pile. Which suggests the recipe was not invented once and passed down. It was invented, and then SHARED.

And then there is the third explanation, the one you will hear in every retreat center from Costa Rica to the Netherlands: that the mushroom makes it for us. That this is a gift, offered by an intelligence we have not learned to recognize, to a species that badly needs it. Hmm.

Of course there is no scientific evidence for that whatsoever, and this book promised to tell you when a beautiful story is probably a myth. Probably a myth but up to you to decide.

But it is a magnificent one, and if you have ever stood in a wet October field at dawn with a basket, you understand exactly why humans keep telling it.

---

## FIELD NOTES: *PSILOCYBE* MUSHROOMS

**WHERE ON EARTH:** Every continent except Antarctica. Concentrated in temperate grassland (*semilanceata*), subtropical pasture (*cubensis*), Pacific Northwest coastal dune (*azurescens*), and Mesoamerican highland (*mexicana*).

**WHERE IT'S LEGAL:** Regulated therapeutic programs in Oregon, Colorado and New Mexico. Decriminalized in roughly thirty US cities and counties. Legal or unregulated in Jamaica, the Bahamas, Brazil, Nepal and Samoa. In the Netherlands the fruiting body is banned but the underground sclerotia, sold as truffles, are legal. Federally Schedule I in the United States regardless of local law.

**HOW LONG WE'VE USED IT:** Rock art evidence suggests six to nine thousand years. Documented ceremonial use in Mesoamerica for at least five hundred years and certainly far longer.

**HOW IT'S OBTAINED:** Wild foraging, overwhelmingly. Home cultivation from spores, which is legal to possess in most states because spores contain no psilocybin. Licensed purchase through Oregon and Colorado service centers. Legal retail truffle shops in Amsterdam. Retreat centers in Jamaica, Mexico, Costa Rica and the Netherlands.

---

### THE STRAIGHT DOPE

The blue bruising test is helpful. It is not proof.

*Galerina marginata* is a small brown mushroom that grows on wood and wood chips, often in the same beds where *Psilocybe cyanescens* grows, sometimes side by side in the same handful. It contains amatoxins, the same class of poison found in the death cap. Amatoxin poisoning destroys the liver. There is no antidote and symptoms do not appear for six to twenty-four hours, by which point the damage is done.

Experienced mycologists are cautious. Wild foraging is the single most dangerous activity described in this entire book, and it has nothing to do with the psychedelic and everything to do with the lookalike in nature. Facts first.

---

### THE RED ONE

Now, through birch forest and deep snow, we travel north.

*Amanita muscaria* is the mushroom your brain draws when someone says the word mushroom. Brilliant scarlet cap. White warts scattered across it like snow. Thick white stem with a ring. It is in every fairy tale, on every garden ornament, in Mario, in Fantasia, on a thousand Victorian Christmas cards. It is arguably the most recognizable organism on the planet. Pure magic.

And here is what almost nobody realizes: it contains no psilocybin at all.

Not a trace. The fly agaric works by an entirely different mechanism using entirely different molecules. Its active compounds are ibotenic acid and muscimol, and rather than fitting the serotonin lock they fit the GABA lock, which is the brain's primary braking system. The result is not a classic psychedelic experience. It is something dreamier and more physically strange. Deep sedation alternating with bursts of energy. Vivid dreaming while awake. Distortions of size, so that objects appear enormous or tiny, an effect called macropsia and micropsia.

Lewis Carroll almost certainly knew about this as he showed us the famous rabbit hole. Alice eats a piece of mushroom and grows and shrinks. Draw your own conclusions.

The fly agaric grows in a grand partnership with the mighty birch, pine and spruce trees across the entire northern hemisphere, from Scotland to Siberia to Colorado. And in Siberia, among the Koryak, the Chukchi and the Kamchadal, it was the sacred sacrament, used exactly as we described in Chapter One, urine recycling and all. Indeed.

One legal oddity worth knowing, because it usually surprises everyone. *Amanita muscaria* is not a controlled substance in the United States. It is federally unscheduled and legal to possess in forty-nine states, with Louisiana the sole exception. You can buy dried caps online right now. Whether you SHOULD is a completely different question, because ibotenic acid is genuinely toxic and the dose response is unpredictable and unpleasant in ways psilocybin simply is not.

The most famous mushroom in the world is also the one most people are wrong about. Facts first.

CHAPTER TWO

# The Red One

*Amanita muscaria: three fruiting bodies*

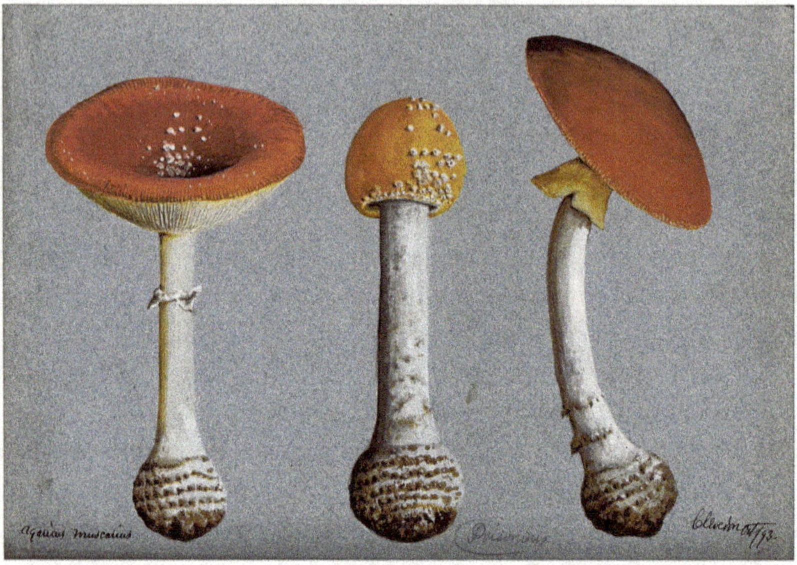

**The most recognizable organism on Earth, and the one most people are wrong about.**

***It contains no psilocybin at all.***

*Watercolor, 1893. Wellcome Collection, CC BY 4.0*

How the World Gets High · TJ Toce

---

## FIELD NOTES: *AMANITA MUSCARIA*

**WHERE ON EARTH:** Circumboreal. Native across the entire northern hemisphere in association with birch, pine and spruce. Introduced and now naturalized in Australia, New Zealand, South Africa and Chile.

**WHERE IT'S LEGAL:** Federally unscheduled in the United States and legal in forty-nine states. Louisiana is the exception. Legal in most of Europe, though the Netherlands and the UK restrict sale for consumption.

**HOW LONG WE'VE USED IT:** Siberian ceremonial use documented by European travelers from the early 1700s and understood to be far older. Possibly, though contentiously, the soma of the Rig Veda, roughly 1500 BCE.

**HOW IT'S OBTAINED:** Wild foraging in northern forest. Legally sold dried, online and in shops, throughout the United States and much of Europe.

---

### THE FUNGUS THAT BURNED

We come now to the dark cousin in the family, and it deserves a careful telling.

*Claviceps purpurea* is a fungus that infects rye and other cereal grains. Instead of a normal grain, the infected head produces a hard, dark, curved growth called a sclerotium, which looks a bit like a rat's claw. Millers did not always notice. Bakers did not always notice. And so, in bad years, whole villages ate bread made from contaminated flour.

What followed was called St. Anthony's Fire, and it came in two forms. Anthony, not Elmo.

The gangrenous form constricted blood vessels so severely that limbs lost circulation and died. Fingers, toes and eventually whole feet blackened and fell away. Sufferers described a sensation of burning.

The convulsive form was worse in a different way. Violent seizures. Muscle spasms. Crawling sensations on the skin. Mania. And hallucinations, terrifying ones, that could last for days.

Epidemics swept through medieval and early modern Europe repeatedly. Tens of thousands died in single outbreaks. The last significant one hit a French village in 1951.

In 1976, a graduate student named Linnda Caporael published a paper proposing that the Salem witch trials of 1692 were touched off by ergot poisoning. The rye harvest that year had been grown in warm, wet conditions that favor the fungus. The afflicted girls' symptoms map onto convulsive ergotism with uncomfortable precision: the fits, the crawling skin, the visions.

It is a compelling theory. It is also disputed, hard, by historians who point out that ergotism should have affected whole households rather than a specific group, and that the social and political fractures in old Salem explain the accusations perfectly well without any fungus at all. The honest answer is that we do not know, and probably never will.

But the story doesn't end here, and what comes next is the most remarkable reversal in this entire book. The alkaloids in ergot, the very compounds that burned and blinded and maddened medieval Europe, are the raw material from which Albert Hofmann built LSD.

The same fungus. One path leads to a village of screaming farmers in 1580. The other leads to a Swiss chemist on a bicycle in 1943, and from there, eventually, to a clinical trial for treatment-resistant depression in 2026. And let's not forget the Summer of Love!

Nature does not hand out gifts and curses in separate boxes. It hands you a molecule, and what happens next is up to us. Isn't that the whole story of this book in one sentence?

---

### FIELD NOTES: ERGOT (*CLAVICEPS PURPUREA*)

**WHERE ON EARTH:** Worldwide wherever rye, wheat, barley and wild grasses grow. Most problematic historically in cool, damp northern Europe.

**WHERE IT'S LEGAL:** The fungus itself is not controlled. Its derivatives are: ergotamine is a prescription migraine medication, ergometrine is used in obstetrics, and LSD is Schedule I.

**HOW LONG WE'VE USED IT:** Accidentally and catastrophically since the beginning of cereal agriculture. Deliberately, in medicine, since midwives used it to control postpartum bleeding centuries ago. Possibly ceremonially at Eleusis, though unproven.

**HOW IT'S OBTAINED:** Not obtained. Avoided. Modern grain cleaning has all but eliminated it from the food supply, which is why you have never heard of anyone getting it these days.

---

### THE NIGHT IN HUAUTLA

We touched on this story in Chapter One. I believe it deserves a closer look, because the details matter.

R. Gordon Wasson was not a scientist. He was a banker, a vice president at J.P. Morgan, and he came to mycology through his wife. Valentina

Pavlovna Wasson was Russian, and on their honeymoon in the Catskills in 1927 she went joyfully into the woods to gather wild mushrooms while he stood on the path, appalled, convinced she was going to poison them both.

That difference fascinated him for the rest of his life. Why do some cultures love mushrooms and others fear them? He coined terms for it: mycophilic and mycophobic. And he and Valentina spent thirty years chasing the question around the world.

It led them to Mexico. On the night of June 29, 1955, in the mountain town of Huautla de Jiménez, Wasson and a photographer named Allan Richardson became almost certainly the first outsiders in the modern era to participate in a Mazatec velada. A sacred, nocturnal healing and divination ceremony.

María Sabina conducted it. The room was dark. She sang for hours, and the singing was not accompaniment, it WAS the ceremony, a river of improvised verse in which she named the saints and the sacred children and the sickness she was working on.

Wasson ate six pairs of mushrooms. He wrote later that he saw geometric patterns give way to palaces and gardens, and that he felt his soul had been scooped out of his body and hung suspended in space. He said that for the first time in his life he understood the word ecstasy in its original sense, which is to stand outside oneself.

He also wrote, and this is the part that gets quoted least yet matters most, that the experience made him doubt whether ordinary waking language was even remotely capable of adequately describing what had happened.

Then he published it in a magazine with a circulation of five and a half million.

We told you in Chapter One what that cost María Sabina. The night itself was real, respectful, and by every account extraordinary. What came after was a stampede. Any honest account of modern psychedelics has to acknowledge the impact of the second one alongside the wonder of the first.

## WHERE WE ARE NOW

Which brings us, remarkably, to the newest venues in the retail world, a perfectly legal shop with a receipt.

**Oregon.** Voters passed Measure 109 in November 2020, and the first licensed psilocybin service center opened in Eugene in June 2023. It is not a dispensary. You do not buy mushrooms and take them home. You book a session, you meet a licensed facilitator beforehand, you consume the psilocybin on site, and the facilitator stays with you for the entire experience, typically six hours. No medical diagnosis is required. The program was substantially amended in March 2026.

**Colorado.** Voters passed Proposition 122 in November 2022, and the legislature built the framework in 2023. Licensed healing centers now operate, and personal use, personal cultivation and sharing without payment are permitted for adults twenty-one and over. Colorado's law covers psilocybin, psilocin, DMT, ibogaine and mescaline, making it the broadest state framework in the country.

**New Mexico.** The Medical Psilocybin Act was signed in April 2025, creating a doctor-supervised medical program rather than Oregon's open-access model. First patient enrollment is targeted for late 2026.

**The Netherlands.** Fresh mushrooms were banned in 2008 after a tourist death. But the ban was written to cover the fruiting body, and *Psilocybe* also forms a dense underground mass called a sclerotium. Sclerotia were not named in the law. And so truffles are sold openly in Amsterdam smart shops to this day, an entire legal industry existing inside a drafting oversight. You have to admire the loophole.

**Jamaica** never scheduled psilocybin at all, which is why a substantial retreat industry has grown up there.

From a rock painting in a green Sahara to a licensed facilitator handing someone a golden chalice in Eugene, Oregon.

Nine thousand years! What a journey. And the mushrooms were so patient the entire time.

CHAPTER TWO

## Four Ways a Country Can Answer

*The legal ladder, from prohibition to regulated access*

**PROHIBITION** — *Most of the world*
Schedule I. No legal access of any kind.

**DEPRIORITIZATION** — *~30 US cities*
Still illegal. Police told to make it a low priority.

**REGULATED ACCESS** — *Oregon · Colorado*
Licensed facilitator, licensed center, supervised session.

**MEDICAL** — *New Mexico*
Qualifying condition, physician involved, prescribed pathway.

MORE ACCESS →

*Federal law is unchanged under every one of these. None of them makes possession federally legal.*

How the World Gets High · TJ Toce

# CHAPTER THREE

## PLANTS

*The biggest room in the house*

> *What is a weed? A plant whose virtues have not yet been discovered.*
>
> **RALPH WALDO EMERSON Fortune of the Republic, 1878**

---

CHAPTER THREE

## The Vine of the Soul

*Banisteriopsis caapi*

One half of the most unlikely recipe on Earth.
*On its own it does almost nothing.*

*Photograph by Anthony Baratier, CC BY-SA 4.0*

How the World Gets High · TJ Toce

If fungi are the oldest room in this house, plants are by far the biggest.

Somewhere between three and four hundred thousand species of plant have been described, and humans have found psychoactive properties in more than a thousand of them. Vines. Leaves. Roots. Bark. Seeds you can buy in a garden center. Spices sitting in your kitchen cabinet right now, this very minute, doing absolutely nothing suspicious.

This chapter travels in three parts to make the story impactful. Flowers and vines first, because that is where the deepest ceremony lives. Then spices, because the story is funnier and stranger than you expect. Then cacti, because the cacti are magnificent and the cacti are in trouble.

Ready? Let's dive into the green.

## PART I: FLOWERS AND VINES

### AYAHUASCA, AND THE MOST UNLIKELY RECIPE ON EARTH

We merely touched on this in Chapter One. Now let's open it all the way up, because the more you understand about how ayahuasca works, the greater the mystery as to how anybody figured it out.

Ayahuasca is a brew. In Quechua the name means roughly "vine of the soul" or "vine of the dead," and it is made by boiling two plants together, sometimes for a full day and a night.

The first plant is *Banisteriopsis caapi*, a thick woody liana that spirals up through the Amazonian canopy. Pounded and boiled, it releases a family of compounds called harmala alkaloids: harmine, harmaline and tetrahydroharmine.

The second plant is usually *Psychotria viridis*, a modest green shrub called chacruna. Its leaves contain DMT, one of the most powerful psychedelic compounds in existence.

Now here is the puzzle, and I want you to really sit with it.

Swallow chacruna leaves alone and nothing happens. Nothing at all. Your body produces an enzyme in the gut and liver called monoamine oxidase (MAO) whose job it is to break down exactly this class of molecule before it can reach your bloodstream. DMT taken by mouth is destroyed on arrival. It is one of the most reliably inactive things you could possibly eat.

Boil the vine alone and you get a mild, nauseating, faintly sedating drink. Interesting, but not transformative.

Put them together, however, and sacred alchemy begins. The harmala alkaloids in the vine are monoamine oxidase inhibitors. They temporarily switch off the enzyme. The gate opens. The DMT sails straight through, reaches the brain, and produces an experience that people spend the rest of their lives trying to describe.

Two plants that work together harmoniously, pun intended, yet they do not grow near each other, look nothing alike, and there is no visible clue whatsoever connecting them.

The Amazon basin holds roughly eighty thousand plant species. Run the arithmetic on finding that specific pair by trial and error and the number gets embarrassing fast.

So how?

Ask a Western ethnobotanist and you will get a careful, respectable answer about many generations of systematic experimentation, about deep observation of animal behavior, and about knowledge accumulated and transmitted across centuries. That is reasonable and may be part of the truth.

Ask a shaman on the Ucayali and the answer takes about four words. The plants told us.

Which explanation is truly stranger?

Archaeology gives us context for how long this has been going on. That ritual bundle from Cueva del Chileno in Bolivia, the one made from three fox snouts, tested positive for both harmine and DMT and dates to roughly a thousand years ago. Ceremonial use is certainly much older than that. We simply do not have older residue.

Today ayahuasca is legal in Peru, Brazil, Ecuador, Colombia and Bolivia, and Peru declared its ceremonial use part of the national cultural heritage in 2008. In the United States, DMT is Schedule I, but two Brazilian churches, the União do Vegetal and Santo Daime, won federal religious exemptions through the courts. The UDV case went all the way to the Supreme Court, which ruled unanimously in their favor in 2006.

A vine, a leaf, and a Supreme Court decision. Most intriguing!

## FIELD NOTES: AYAHUASCA

**WHERE ON EARTH:** The Amazon basin, across Peru, Brazil, Ecuador, Colombia and Bolivia. *B. caapi* is now also cultivated in Hawaii, Costa Rica and parts of Southeast Asia.

**WHERE IT'S LEGAL:** Legal for ceremonial use in Peru, Brazil, Ecuador, Colombia and Bolivia. Costa Rica permits it in practice. In the United States, permitted only for members of the UDV and Santo Daime under federal religious exemption. Colorado's Natural Medicine Act covers DMT. Illegal in most of Europe, with the Netherlands having reversed an earlier tolerance.

**HOW LONG WE'VE USED IT:** Chemically confirmed at approximately one thousand years. Almost certainly several thousand.

**HOW IT'S OBTAINED:** Prepared by a trained shaman or ceremonial leader, typically boiled over one to three days. Obtained by traveling to a retreat center in South America, by attending a legally recognized church ceremony, or through underground ceremonies that operate in most major Western cities and carry real legal and safety risk.

---

## THE SEEDS IN THE GARDEN CENTER

Here is something that delights me every single time I think about it.

Walk into any garden center in America. Find the flower seed rack. Look for morning glory, the variety called Heavenly Blue, or Pearly Gates, or Flying Saucers.

Those seeds contain lysergic acid amide. LSA. A close chemical relative of LSD.

The Aztecs and the Zapotec knew this perfectly well. Their name for the seeds of the related species *Turbina corymbosa* was ololiuqui, and Spanish chroniclers of the sixteenth century recorded its ceremonial use with the same alarm they reserved for teonanácatl. Sahagún wrote it down. The Inquisition banned it. It went into the hills alongside the mushrooms and never fully left.

Then, in 1960, Albert Hofmann (of course it's Hofmann again, this man is everywhere) analyzed ololiuqui seeds and found ergine, or LSA, which is structurally a cousin of the compound he had synthesized in Basel two decades earlier.

He was, by his own account, astonished. He had built LSD from a fungus growing on rye in Europe. And here was a chemically related molecule in a flowering vine in Mexico that indigenous people had been using ceremonially for centuries.

Hawaiian baby woodrose, *Argyreia nervosa*, found mainly on Kaua'i and Maui, carries the same compounds in higher concentration.

Keep in mind LSA is not LSD. It is considerably less potent, and the seeds also carry compounds that cause real nausea, vasoconstriction and misery. Commercial seeds are frequently treated with fungicide. The experience is widely described as unpleasant. This is a wonderful piece of botanical history and a poor idea in practice.

But the fact remains, and it is glorious: one of the sacred plants of Mesoamerica is sitting on a spinner rack at your local nursery for $2.79 in a paper envelope, next to the marigolds.

---

## THE SAGE THAT BREAKS THE RULES

*Salvia divinorum.* I love saying that name. A member of the mint family growing in a small, cloud-wrapped stretch of the Sierra Mazateca in Oaxaca, the same mountains as Huautla de Jiménez, and as far as anyone can tell it does not reproduce in the wild. It sets almost no viable seed. Every plant is a cutting of another plant, propagated by hand.

Which means somebody has been deliberately keeping this species alive, one cutting at a time, for a very long time.

The Mazatec call it *ska maría pastora*, the leaves of Mary the Shepherdess, and they use it in the same tradition as the mushrooms, generally when mushrooms are out of season.

And here is why I am giving salvia its own section. Every other psychedelic in this book works on the serotonin system. LSD, psilocybin, mescaline, DMT. All of them fit the 5-HT2A lock we described in Chapter Two. It is the great unifying mechanism of the entire field.

Salvinorin A, the active compound in Salvia divinorum, does not touch it.

Salvinorin A works on the kappa-opioid receptor, an entirely different system with an entirely different function. It is also the only known naturally occurring psychoactive of its kind that contains no nitrogen atom, which for a chemist is quite surprising.

Users describe brief, extremely intense experiences of becoming an object, of physical space folding, of being pulled sideways. It only lasts minutes rather than hours and can be most unpleasant. There is a long historical record of it being used specifically to **diagnose the root causes of illnesses**, though this occurred within traditional indigenous shamanic rituals rather than modern self-diagnosis.

Legally it is a patchwork. Salvia is not federally scheduled in the United States, but roughly half the states have banned or restricted it individually, so it is legal in my home state of Maine and a felony a little way south in Delaware.

One species. One mountain range. One receptor nobody else uses. Nature keeps a few cards face down. Facts first.

---

## THE ROOT THAT INTERRUPTS ADDICTION

Now we cross the Atlantic to West Central Africa, to the rainforests of Gabon and Cameroon, and to a shrub called *Tabernanthe iboga*.

The Bwiti tradition uses the powdered root bark as its central sacrament. In the initiation ceremony, a candidate consumes a very large dose and remains in a visionary state for a day or more, during which they are said to meet their ancestors and receive their name. It is not recreational and nobody has ever pretended otherwise. Bwiti practitioners describe it as the most serious thing a person does.

Fast forward to 1962, when a nineteen-year-old heroin user in New York named Howard Lotsof took ibogaine, the primary alkaloid, expecting a psychedelic experience.

Something else happened. When it wore off, thirty-some hours later, he had no withdrawal symptoms and no craving. He described walking outside and noticing, with astonishment, that he simply did not want heroin anymore.

He spent the next thirty years of his life trying to get anyone in authority to take this seriously.

The research since has been genuinely striking. Ibogaine appears to interrupt opioid dependence in a way nothing else does, compressing withdrawal into a single session and substantially reducing craving afterward. Studies of veterans treated at clinics in Mexico have reported large improvements in PTSD, depression and cognitive function.

And now for the big but, because this book promised honesty over enthusiasm.

Ibogaine can stop your heart. It affects cardiac rhythm in a way that has killed people, particularly those with existing heart conditions or those who took it alongside opioids. Fatalities have occurred at unsupervised clinics. This is not a substance anyone should approach outside a setting with cardiac monitoring and a physician present.

That said, it is one of the most promising things in this entire field. In 2025 Texas approved substantial state funding for ibogaine clinical development, which is not a sentence anyone expected to write about Texas. More to come.

---

## FIELD NOTES: IBOGA

**WHERE ON EARTH:** Native to the rainforest understory of Gabon, Cameroon and the Republic of the Congo. Now cultivated in Costa Rica, Mexico and Brazil to reduce pressure on wild populations, which are threatened.

**WHERE IT'S LEGAL:** Legal and nationally protected in Gabon, where Bwiti is a recognized religion. Ibogaine is Schedule I in the United States, though covered by Colorado's Natural Medicine Act and the subject of Texas state research funding. Unregulated in Mexico, Costa Rica, Brazil, New Zealand and South Africa, where treatment clinics operate.

**HOW LONG WE'VE USED IT:** Bwiti practice is documented to the nineteenth century and understood by practitioners to be far older.

**HOW IT'S OBTAINED:** Ceremonial initiation in Gabon. Treatment clinics in Mexico, Costa Rica, New Zealand and South Africa. Clinical trials now beginning in the United States.

---

## THE ONES THAT ARE NOT INVITED

A quick, serious word before we move on, and then we get back to the fun.

*Datura* and *Brugmansia*, the jimsonweeds and angel's trumpets, are beautiful plants with enormous trumpet flowers that grow all over the world, including in a great many suburban gardens.

They are not psychedelics. They are called deliriants, and the difference is enormous.

A psychedelic alters your perception while leaving you aware that your perception is being altered. You know you took something. Deliriants do not do this. They produce fully convincing hallucinations that are indistinguishable from reality, along with amnesia, wild confusion, dangerously elevated heart rate, and body temperatures that can climb to lethal levels. People on datura have walked into traffic and have died.

Indigenous traditions across the Americas, India and Europe use these plants, always in tiny quantities, always under close supervision, and always with a healthy respect that modern thrill-seekers have not inherited.

They are in this book for completeness and because you may encounter them growing in a neighbor's yard.

Please leave them there. Facts first.

# PART II: SPICES

Now for something considerably lighter, and a good deal odder.

Open your kitchen cabinet. Somewhere in there, probably behind the Ceylon cinnamon, is a small jar of nutmeg. If you're a serious cook, a few seeds and a grater will be there instead.

That jar and those seeds contain myristicin.

## THE SPICE THAT WENT TO PRISON

Nutmeg is the seed of *Myristica fragrans*, a tree native to the Banda Islands in eastern Indonesia, which for centuries were the only place on Earth it grew. Wars were fought over these islands. The Dutch traded Manhattan to the English partly to secure them. Nutmeg was once worth more than its weight in gold.

We are talking about it here because in large quantities, it is indeed psychoactive.

Myristicin, along with related compounds elemicin and safrole, converts in the body into substances chemically related to mescaline and amphetamine. Consume enough and you get hallucinations, euphoria, dissociation and time distortion, lasting up to two full days.

This has been known for a long time. Medieval European texts mention it. And in twentieth century American prisons, where nutmeg was available in the kitchen and almost nothing else was, it became something close to institutional folklore.

Malcolm X wrote about it in his autobiography. In Charlestown State Prison, he described inmates buying penny matchboxes of nutmeg from kitchen workers, and said that a box "had the kick of three or four reefers."

So why is nutmeg not a global phenomenon?

Because it is dreadful. Truly, comprehensively dreadful. The dose required is roughly two whole nutmegs, which tastes like eating a dirty candle. Onset takes up to six hours, so you cannot tell whether it is working. And the side effects are severe nausea, vomiting, racing heart, crushing headache, dry mouth, dizziness, and a hangover lasting several days.

Nutmeg is nature's practical joke. It absolutely works. Nobody ever does it twice.

## THE FAMILY TREE IN YOUR CABINET

Here is something genuinely elegant.

Myristicin, elemicin and safrole all belong to a chemical family called phenylpropenes, and they turn up all over the spice rack. Safrole is the primary component of sassafras oil, which is what gave old-fashioned root beer its flavor before the FDA banned it in 1960. Elemicin appears in nutmeg and in certain parsley oils. Dill, parsley, black pepper and carrot seed all carry small quantities of related compounds.

And safrole, through a few chemical steps, is a precursor to MDMA. Which is why sassafras oil is now a federally watched chemical, and why the DEA pays attention to shipments of it.

Root beer and ecstasy, sharing a great-grandparent. A spice rack keeping curious secrets.

## THE DESERT SHRUB THAT DOES THE VINE'S JOB

*Peganum harmala*, Syrian rue, is a scrubby, unremarkable plant that grows across a huge swath of the Middle East, North Africa, Central Asia and, thanks to an unfortunate introduction, the American Southwest.

Its seeds contain harmine and harmaline. The very same monoamine oxidase inhibitors found in the ayahuasca vine we discussed earlier.

Which means a plant growing wild in Iran, Turkey, Morocco and New Mexico performs the exact chemical function of a liana growing five thousand miles away in the Peruvian Amazon. Two completely unrelated plants, on opposite sides of the planet, arriving independently at the same solution.

Are you seeing the pattern yet? If not, you will because it keeps happening.

Syrian rue has been burned as incense across Central Asia and the Middle East for millennia. It is used against the evil eye in Türkiye to this day. And some scholars, including our friend Wasson at one point, argued that it, and not the fly agaric, was the true soma of the Rig Veda.

## AND A FEW MYTHS, PUT TO REST

While we are in the kitchen, let's clean up three things everybody believes.

**1. Absinthe does not make you hallucinate.** The wormwood compound thujone is present in quantities far too small to do anything. What absinthe has is an alcohol content of up to seventy-five percent. Nineteenth century Parisian artists were not tripping. They were extremely, magnificently drunk.

**2. Kava is not a psychedelic.** The Pacific Island root drink is genuinely wonderful, socially central across Fiji, Vanuatu, Tonga and Samoa, and pharmacologically it is an anxiolytic and muscle relaxant. It calms and relaxes. It does not open doors other than to a few pupus.

**3. Betel nut is not a psychedelic either.** Chewed by hundreds of millions of people across South and Southeast Asia, it is a stimulant, closer to strong coffee with a nicotine edge. It is also a serious carcinogen causing very nasty cancers.

Getting these right matters. A book that says everything is psychedelic is a book that means nothing when it says something is. Facts first.

# PART III: CACTI

CHAPTER THREE

## Peyote

*Lophophora williamsii, one specimen in flower*

A cactus that grows half an inch a decade and has been used for 5,700 years.
*Most of the plant is underground.*

*N. L. Britton and J. N. Rose, The Cactaceae, vol. III, Plate IX, 1922. Public domain.*

How the World Gets High · TJ Toce

And now we come to my second favorite behind fungi.

There is something special about the cacti. A peyote button sitting in the Chihuahuan Desert has been sitting there, in that exact spot, for perhaps thirty years, growing at a pace that makes glaciers look impatient, waiting.

## PEYOTE

*Lophophora williamsii* is a small, spineless, blue-green cactus that grows flush with the ground across the Chihuahuan Desert, from south Texas down into San Luis Potosí. Most of the plant is underground. What you see is a button a few inches across, divided into soft ribbed sections, with a tuft of white fuzz at the center.

It contains mescaline, the first classical psychedelic ever isolated by science, extracted by the German chemist Arthur Heffter in 1897 and synthesized by Ernst Späth in 1919, decades before anyone had heard of LSD. Another star during the Summer of Love.

Dried peyote buttons recovered from the Shumla Caves in south Texas have been radiocarbon dated to approximately fifty-seven hundred years ago. Let that land. Older than the Great Pyramid. Older than Stonehenge. Older than the written word.

Peyote traveled north across the plains in the eighteenth and nineteenth centuries, reaching the Comanche, Kiowa, Apache and eventually the Lakota and beyond, precisely during the decades when those nations were being destroyed. The Comanche leader Quanah Parker became one of its most influential advocates after peyote was used to treat him following a serious injury.

In 1918, in Oklahoma, the Native American Church (NAC) formally incorporated.

And in 1994, amendments to the American Indian Religious Freedom Act made federal protection explicit: members of federally recognized tribes may possess, transport and use peyote in bona fide ceremony, anywhere in the United States, regardless of state law.

Read that again, because it is astonishing. In a country where mescaline is a Schedule I drug, several hundred thousand people now have a federal right to consume it in ceremony and have had that right for over thirty years. A good thing!

**But peyote is in serious trouble.**

The plant grows unbelievably slowly. Legal harvest in the United States is confined to a small strip of south Texas and conducted by a handful of licensed dealers called peyoteros, whose numbers have dwindled to a very few. Wild populations have been depleted by overharvesting, ranch land conversion, root plowing, and increasingly by non-Native seekers who read a book and went looking.

Native American Church leaders have been remarkably clear about this. Many have asked, publicly and repeatedly, that non-Native people leave peyote alone entirely and use San Pedro or synthetic mescaline instead.

That request is easy to honor and critical to the survival of this magical cactus. Please honor it.

---

### FIELD NOTES: PEYOTE

**WHERE ON EARTH:** Chihuahuan Desert. A narrow band of south Texas along the Rio Grande, and northern Mexico through Coahuila, Nuevo León, Tamaulipas, Zacatecas and San Luis Potosí. Wild populations declining sharply.

**WHERE IT'S LEGAL:** In the United States, federally protected for members of federally recognized tribes in bona fide Native American Church ceremony, under the 1994 AIRFA amendments. Schedule I for everyone else. Protected in Mexico for Huichol (Wixárika) ceremonial use. Colorado's Natural Medicine Act covers mescaline but deliberately excludes peyote at the explicit request of Native leaders.

**HOW LONG WE'VE USED IT:** Radiocarbon confirmed at approximately 5,700 years. The oldest chemically dated psychedelic use anywhere on Earth.

**HOW IT'S OBTAINED:** Wild-harvested in south Texas by a small number of federally licensed peyoteros and distributed to Native American Church congregations. Conservation cultivation programs are now underway. There is no legal path for non-Native access, and Native leaders have asked that none be sought.

---

### SAN PEDRO, THE GENEROUS ONE

Now south, to the Andes, and to a completely different temperament.

*Trichocereus pachanoi*, called San Pedro or huachuma, is a tall columnar cactus that grows on the western slopes of the Andes through Peru, Ecuador and Bolivia. Where peyote is tiny and slow, San Pedro is fast and abundant, capable of adding a foot or more a year and reaching twenty feet tall.

It also contains abundant mescaline, exactly the same compound as peyote.

And the ceremonial record is extraordinary. At Chavín de Huántar, a temple complex in the Peruvian highlands built around 1200 BCE, there is a carved stone relief showing a figure holding a San Pedro cactus. The site includes underground galleries designed with astonishing acoustic properties, and archaeologists have recovered snuff paraphernalia and cactus remains there.

More than three thousand years of continuous documented use in the Andes, and the tradition never fully stopped. Huachumeros still practice on the north coast of Peru today.

The name San Pedro, Saint Peter, comes from the colonial period, when practitioners explained to Spanish priests that just as Saint Peter holds the keys to heaven, the cactus opens the gate. That is either brilliant theological diplomacy or a very good joke, and I choose to believe it is both.

Practitioners describe the San Pedro experience as gentler and more grounded than peyote, longer in duration, often twelve to fourteen hours, and more oriented toward the natural world and the body than toward visionary spectacle.

**And here is the legal curiosity.** Mescaline is Schedule I in the United States. But the *cactus itself* is legal to buy, sell, own and grow as an ornamental plant. It is in nurseries. It is on Amazon. It is in landscaping across the American Southwest. The only thing illegal is preparing it for consumption.

So a plant containing a Schedule I substance sits legally in thousands of American front yards, quietly minding its own business, while its tiny cousin four hundred miles away is federally protected and nearly extinct.

The law is a strange and curious animal.

CHAPTER THREE

# San Pedro

*Echinopsis pachanoi · huachuma*

Where peyote is tiny and slow, San Pedro is fast and abundant, *capable of adding a foot a year and reaching twenty feet.*

*Photograph by Carlo Brescia, CC BY-SA 4.0*

How the World Gets High · TJ Toce

---

## FIELD NOTES: SAN PEDRO (HUACHUMA)

**WHERE ON EARTH:** Andean slopes of Peru, Ecuador and Bolivia, between roughly 6,000 and 9,800 feet. Widely cultivated as an ornamental worldwide, including throughout the American Southwest, Australia and the Mediterranean.

**WHERE IT'S LEGAL:** Ceremonial use legal in Peru, Ecuador and Bolivia. In the United States the living plant is legal to grow and sell as an ornamental, while extracting or preparing it for consumption is a Schedule I offense. Colorado's Natural Medicine Act covers mescaline from non-peyote sources.

**HOW LONG WE'VE USED IT:** Over 3,000 years, with iconographic evidence at Chavín de Huántar dating to approximately 1200 BCE.

**HOW IT'S OBTAINED:** Ceremony with a huachumero in Peru and Ecuador. Purchased as a live ornamental plant in nurseries and online in the United States, where the intent behind the purchase determines the legality.

---

CHAPTER THREE

## Two Cacti, One Molecule

*Why the choice between them matters*

| | PEYOTE<br>*2 inches across, after 30 years* | SAN PEDRO<br>*up to 20 ft, a foot a year* |
|---|---|---|
| SPECIES | Lophophora williamsii | Trichocereus pachanoi |
| COMMON NAME | Peyote | San Pedro · Huachuma |
| WHERE | Chihuahuan Desert | Andean slopes, 6,000–9,800 ft |
| GROWTH RATE | half an inch per decade | up to a foot per year |
| DOCUMENTED USE | 5,700 years ago | 3,000+ years ago |
| LEGAL STATUS | Federally protected for NAC members. Schedule I for everyone else. | Legal to buy and grow as an ornamental. Preparing it is a Schedule I offense. |
| CONSERVATION | Wild populations in serious decline | Abundant and widely cultivated |

*Same primary compound. Same mescaline. Two completely different plants, and two completely different obligations.*

*Native American Church leaders have asked that non-Native seekers use San Pedro and leave peyote alone.*

How the World Gets High · TJ Toce

### HUXLEY AT THE FLOWER VASE

One last snippet before we leave the green.

In May 1953, in Los Angeles, a fifty-eight-year-old English novelist named Aldous Huxley, whose books I adore, swallowed four tenths of a gram of mescaline dissolved in a glass of water, sat down, and waited.

He was already famous. *Brave New World* had been published twenty-one years earlier. He was also nearly blind, having lost most of his sight to an illness at sixteen, which makes what he wrote next even more remarkable.

He looked at a small glass vase holding three flowers. And he saw, he said, "what Adam had seen on the morning of his creation, the miracle, moment by moment, of naked existence."

He looked at the folds of his grey flannel trousers and found them so beautiful and so complex that he understood, for the first time, what draperies in Old Master paintings were actually about.

*The Doors of Perception* was published in 1954. It is barely sixty pages. And it did something no scientific paper had managed, which was to give ordinary English-speaking readers a vocabulary for this experience that did not sound clinical and did not sound insane. I highly recommend reading it.

The title comes from William Blake: *If the doors of perception were cleansed, every thing would appear to man as it is, Infinite.*

A decade later a young band in Los Angeles read that line and named themselves after it. I drive past Jim Morrison's old house in Laurel Canyon most weekends when I am at my California home.

Every conversation the West has had about psychedelics since 1954 has been shaped by a nearly blind novelist looking very carefully at three flowers in a vase.

That is the power of paying attention. And attention, in the end, may be what this entire subject is really about. Think about it.

CHAPTER THREE

# The Plant Belt

*Native ranges of the six major psychoactive plants*

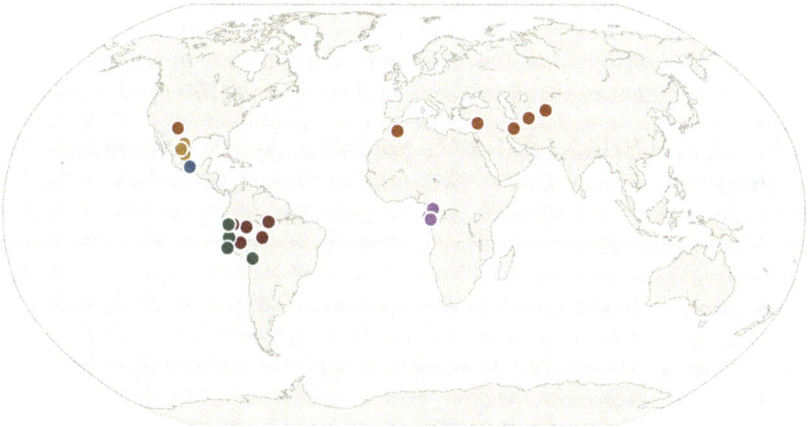

- **Ayahuasca**
  *B. caapi + P. viridis, Amazon basin*
- **San Pedro**
  *T. pachanoi, Andean slopes*
- **Salvia divinorum**
  *Sierra Mazateca, Oaxaca*
- **Peyote**
  *L. williamsii, Chihuahuan Desert*
- **Iboga**
  *T. iboga, Gabon and Cameroon*
- **Syrian rue**
  *P. harmala, arid Old World and SW USA*

How the World Gets High · TJ Toce

# CHAPTER FOUR

## AMPHIBIANS

*The one nobody believes until they look it up*

> *Toad, that under cold stone / Days and nights has thirty-one / Sweltered venom sleeping got, / Boil thou first i' th' charmèd pot.*
>
> **WILLIAM SHAKESPEARE Macbeth, Act IV Scene i, c. 1606**

---

---

We have covered fungi. We have covered plants. Both of those, most people can accept.

Now we arrive at the toad, and this is the point in every conversation about this subject where somebody laughs and says "come on."

Come on indeed. Let's go meet him.

---

### THE DESERT AND THE RAIN

For most of the year, *Incilius alvarius* is not doing anything at all. He is underground, in a burrow in the Sonoran Desert, waiting out temperatures that regularly clear a hundred and ten degrees.

Then, in late June or early July, the monsoon arrives. Rain hammers the desert floor. And within hours, thousands of large, olive-green toads emerge from the ground, make their way to temporary pools, and begin to call.

He is a serious animal. Adults reach seven inches, which makes him the largest native toad in the United States. Smooth skin, warty flanks, and behind each eye a pair of swollen kidney-shaped organs called parotoid glands, with more scattered along the legs.

Those glands are a defense system. Squeeze them and they produce a thick white secretion that has, over evolutionary time, taught nearly every predator to leave this toad alone.

That secretion contains bufotenine and a range of cardiotoxic compounds called bufadienolides. And it contains something else, in remarkable quantity.

5-methoxy-N,N-dimethyltryptamine. 5-MeO-DMT.

*Incilius alvarius* is the only animal on Earth known to produce this compound in significant amounts.

---

## WHAT IT ACTUALLY DOES

5-MeO-DMT is not ayahuasca. It is not psilocybin. It is not LSD. Users are almost unanimous that comparing it to those is like comparing a hurricane to weather.

The secretion is dried into a crystalline material and vaporized. Onset is under thirty seconds. The peak lasts perhaps five to fifteen minutes. The entire experience is over in half an hour.

And in those few minutes, people report something that recurs so consistently across accounts that researchers have taken serious note of it: the complete dissolution of the sense of being a separate self. Not visions. Not patterns. Not scenery. The reported experience is of the observer itself disappearing into undifferentiated unity, with no remaining sense of a person who is having the experience.

Survey research from Johns Hopkins has found that a very large majority of 5-MeO-DMT users rate the session among the most personally meaningful experiences of their entire lives, with ratings that meet or exceed those reported for high-dose psilocybin.

Five minutes. It is the shortest and, by many accounts, the most total experience in this entire book.

Can you imagine? Neither can anyone who hasn't, which is rather the point.

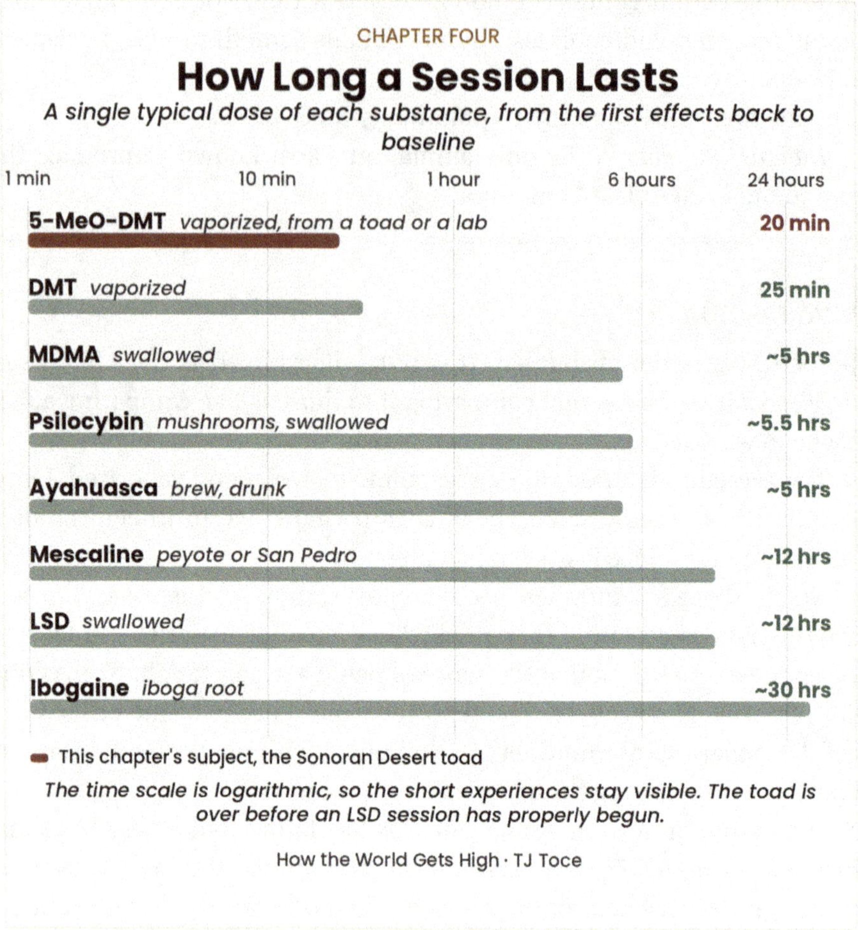


## NO, YOU CANNOT LICK IT

Let's kill this one properly, because it will not die on its own.

The phrase "licking toads" entered American popular culture in the 1980s and has never left. It has been in newspaper headlines, in *The Simpsons*, in a hundred stand-up routines. National park services have issued official statements about it.

Here is the reality.

The secretion contains bufadienolides, which are cardiac glycosides. They act on the heart in a manner similar to digitalis. Swallowing them can cause vomiting, seizures, cardiac arrhythmia and death. Dogs that mouth these toads in Arizona backyards die with genuinely tragic regularity, and every veterinarian in Tucson knows the protocol by heart.

Furthermore, raw 5-MeO-DMT taken orally is largely inactivated by the gut, the same way DMT is. So licking a toad would deliver you a mouthful of heart poison and essentially none of the compound anyone is after.

What is actually done is that the secretion is expressed onto a glass surface, dried into a crust, scraped off and vaporized. The myth is more famous than the reality. Facts first.

---

## THE TOAD IS PAYING FOR OUR ENTHUSIASM

And now the part that matters most in this chapter.

Demand for this experience has exploded. Retreats offering toad ceremonies operate in Mexico, the United States and across Europe. Celebrities have talked about it on podcasts. And the animal supplying it lives in one specific desert, emerges above ground for a few weeks a year, and is being collected, handled and milked at a scale it was never built to withstand.

The problems compound. Toads are being taken from the wild and trafficked. Repeated milking stresses the animal. Some collectors do not return them and their habitat is disappearing across the range. New Mexico already lists the species as threatened. Populations in parts of California appear to be gone entirely.

There is also a cultural question that deserves a straight answer. Toad ceremonies are frequently marketed as ancient indigenous practice. The archaeological record shows toad iconography in Olmec and Maya art, and *remains of Bufo marinus, since reclassified as Rhinella marina, have been found at Maya sites*, so amphibians unquestionably mattered in Mesoamerican cosmology. But the specific practice of vaporizing dried *Incilius alvarius* secretion appears to date to the 1980s, following a 1984 pamphlet by an author writing as Albert Most.

That does not make it less powerful. It does make the marketing dishonest.

**Here is the good news, and it is genuinely good.** 5-MeO-DMT can be synthesized in a laboratory. The synthetic molecule is chemically identical. It contains none of the cardiotoxic bufadienolides, so it is arguably safer. It is consistent in dose, which the toad never is. And it requires no animal at all.

Nearly every serious researcher and a growing number of practitioners have now moved to synthetic exclusively. Conservation organizations in Arizona and Sonora are asking everyone else to do the same.

You can have the entire experience and leave the toad in his burrow, waiting for the monsoon.

That seems like an easy call, doesn't it?

---

### FIELD NOTES: SONORAN DESERT TOAD

**WHERE ON EARTH:** The Sonoran Desert. Southern Arizona, extreme southwestern New Mexico, and the Mexican states of Sonora and Sinaloa. Historically present in southeastern California, where it is now believed extirpated.

**WHERE IT'S LEGAL:** 5-MeO-DMT is Schedule I in the United States. The live toad is legal to possess in Arizona with a fishing license but illegal to collect in New Mexico, where the species is listed as threatened. Colorado's Natural Medicine Act covers DMT. Unregulated in Mexico, where most retreats operate.

**HOW LONG WE'VE USED IT:** The modern practice of vaporizing dried secretion dates to the 1980s. Broader amphibian symbolism in Mesoamerican ceremony goes back millennia, but the direct link is unproven.

**HOW IT'S OBTAINED:** Milked from wild-caught toads and dried, which is the practice this book asks you not to support. Or synthesized in a laboratory, which is identical, safer, more consistent and harmless to the animal.

---

### THE STRAIGHT DOPE

5-MeO-DMT is the substance in this book with the smallest margin between profound and dangerous.

It carries genuine risk of respiratory depression and loss of consciousness. There have been deaths. It is extremely dangerous in combination with monoamine oxidase inhibitors, which includes both ayahuasca preparations and a number of common antidepressants, and that combination has killed people.

Nobody should be anywhere near this without a sober, trained sitter physically present for the entire session. Facts first.

---

## KAMBO, WHICH IS NOT THIS

Now, a necessary detour, because kambo gets swept into this conversation constantly and it does not belong.

*Phyllomedusa bicolor* is the giant monkey frog, a bright green tree frog with enormous eyes that lives in the Amazon canopy across Brazil, Peru, Colombia and Bolivia. The Matsés, the Katukina and other Amazonian peoples use its skin secretion in a practice called kambo.

Small points are burned into the skin, usually on the upper arm or leg, and the dried secretion is applied directly to the burns. What follows is fast and unpleasant: intense flushing, a racing heart, swelling of the face, and violent vomiting, lasting perhaps twenty to forty minutes.

It is used traditionally to sharpen hunting ability, to clear what practitioners call *panema*, a kind of bad luck or heaviness, and as a general tonic before the hunt.

**Kambo is not a psychedelic.** There is no hallucination, no altered perception, no visionary content whatsoever. You are entirely yourself the whole time, feeling extremely unwell.

Deaths have been reported, generally linked to excessive water intake before the session, to underlying heart conditions, or to sessions run by untrained practitioners.

It is here because you will hear it mentioned in the same breath as ayahuasca and toad, and you should know it is a completely different thing. Facts first.

---

## FIELD NOTES: KAMBO

**WHERE ON EARTH:** Amazon rainforest canopy across Brazil, Peru, Colombia, Bolivia, Venezuela and the Guianas.

**WHERE IT'S LEGAL:** The secretion is not a controlled substance in the United States, though the FDA has issued warnings against its sale as a health product. Legal in Brazil and Peru. Banned for practitioner use in Italy following a fatality. Regulated in Australia.

**HOW LONG WE'VE USED IT:** Documented among the Matsés and neighboring peoples since at least the early twentieth century, with oral tradition indicating far longer.

**HOW IT'S OBTAINED:** The frog is gently secured and its secretion collected on a wooden stick, then dried. Traditionally the frog is released unharmed, and Matsés practice holds that mistreating the frog brings misfortune.

---

## TOADS IN STONE

Let's finish where the archaeology actually is, rather than where the marketing claims it is.

Amphibians appear all over Mesoamerican art. The Olmec, the earliest major civilization in the region, carved toad effigies. Maya ceramics show toads. At the Olmec site of San Lorenzo, archaeologists recovered the skeletal remains of a large number of *Bufo marinus*, the cane toad, in a context that looks deliberate rather than accidental.

The cane toad is a different animal from our Sonoran friend. It contains bufotenine and a heavy load of cardiotoxins, and importantly it does NOT contain meaningful quantities of 5-MeO-DMT. Attempts to use it as a psychedelic have poisoned people.

So what were the Olmec doing with all those toads?

The honest answer is that we do not know. The toad is a powerful symbol in its own right without any chemistry involved. It buries itself in the dry earth and appears to die. Then the rains come, and it emerges. In an agricultural society utterly dependent on the arrival of the monsoon, an animal that resurrects itself when the rain returns is a natural emblem of renewal, fertility and the turning of the year.

Maybe the toads were medicine. Maybe they were metaphor. Maybe, as is so often true, they were both at once.

That is where the evidence stops, and this book stops with it.

CHAPTER FOUR

# The Sonoran Desert Toad

*Range of Incilius alvarius, the only animal known to produce 5-MeO-DMT in quantity*

- **Current range**
- **Historic range, now believed extirpated**

**Bufo alvarius**

*Mary C. Dickerson, The Frog Book, 1906*
*Plate V, figure 4 · public domain*

Up to seven inches long, the largest native toad in the United States. The gland behind the eye is the one that produces 5-MeO-DMT.

How the World Gets High · TJ Toce

# CHAPTER FIVE

## FISH

*A lovely dinner, and then the strangest thirty-six hours of your life*

*Nature is nowhere greater than in the smallest things.*

**PLINY THE ELDER Natural History, c. 77 AD**

---

---

This is the shortest chapter in the book and quite possibly the best one to bring up at a dinner party.

In 2006, a forty-year-old man was on holiday on the French Riviera. He had dinner at a world-class restaurant in Saint-Tropez, and he ordered the fish.

Two hours later he began to feel very ill. His vision blurred. Then, driving home, he started seeing things. Screaming animals appeared around his car. He pulled over. The hallucinations continued through the night and did not end for THIRTY-SIX HOURS.

He went to the hospital, where doctors, understandably, wanted to know what he had taken. Nothing, he said. I had the fish special.

He was telling the truth.

---

### THE WORD OF THE DAY

The condition is called **ichthyoallyeinotoxism**.

Say it out loud. Go on, it is worth the effort. *Ick-thee-oh-ah-lee-ee-no-tox-ism*. From the Greek: *ichthys* for fish, *allyeino* for to make crazy, *tox-ikon* for poison. For the linguists!

The fish that makes you crazy.

That word alone is worth the price of this book, and you now know something that roughly ninety-nine percent of the people you will ever meet do not.

---

## THE FISH ITSELF

*Sarpa salpa* is a sea bream. It is about a foot long, silver, with a row of neat golden stripes running head to tail, and it looks like exactly what it is, which is a perfectly ordinary Mediterranean food fish.

It lives along the coasts of the Mediterranean, down the eastern Atlantic from the Bay of Biscay to South Africa, and around the Cape into the western Indian Ocean. It grazes in shallow water on algae and seagrass beds. It shoals. It is caught, sold and eaten routinely from Barcelona to Beirut, and the overwhelming majority of the time, nothing happens at all.

But not always.

The 2006 Saint-Tropez case is the most cited, and it has a companion. In 1994 a ninety-year-old man ate *Sarpa salpa* in the south of France and spent the next two days tormented by auditory hallucinations of human screams and birdsong, and by visions he found so disturbing that he refused to describe them fully to his doctors. Awful!

Note that detail, because it matters. This is not a pleasant experience. Nobody is seeking this out. Every documented case reads as an unwanted, frightening episode that arrived with the check.

## THE ROMANS KNEW

Consider this. The Romans apparently ate it deliberately.

Pliny the Elder, in his *Natural History*, mentions *Sarpa salpa* and its effects. Later Roman sources indicate the fish was consumed as a recreational intoxicant among people wealthy enough to be adventurous with dinner.

The Arabic name for the fish, recorded in North Africa, translates as *the fish that makes dreams*.

So this has been known, named and occasionally sought after for at least two thousand years, and yet almost nobody alive today has heard of it. How does that happen? How does a hallucinogenic fish stay a secret in plain sight, in the middle of the Mediterranean, for two millennia?

Mostly because it is unreliable, quite unpleasant and rare. Which is a shame for the storyteller and probably a mercy for everyone else.

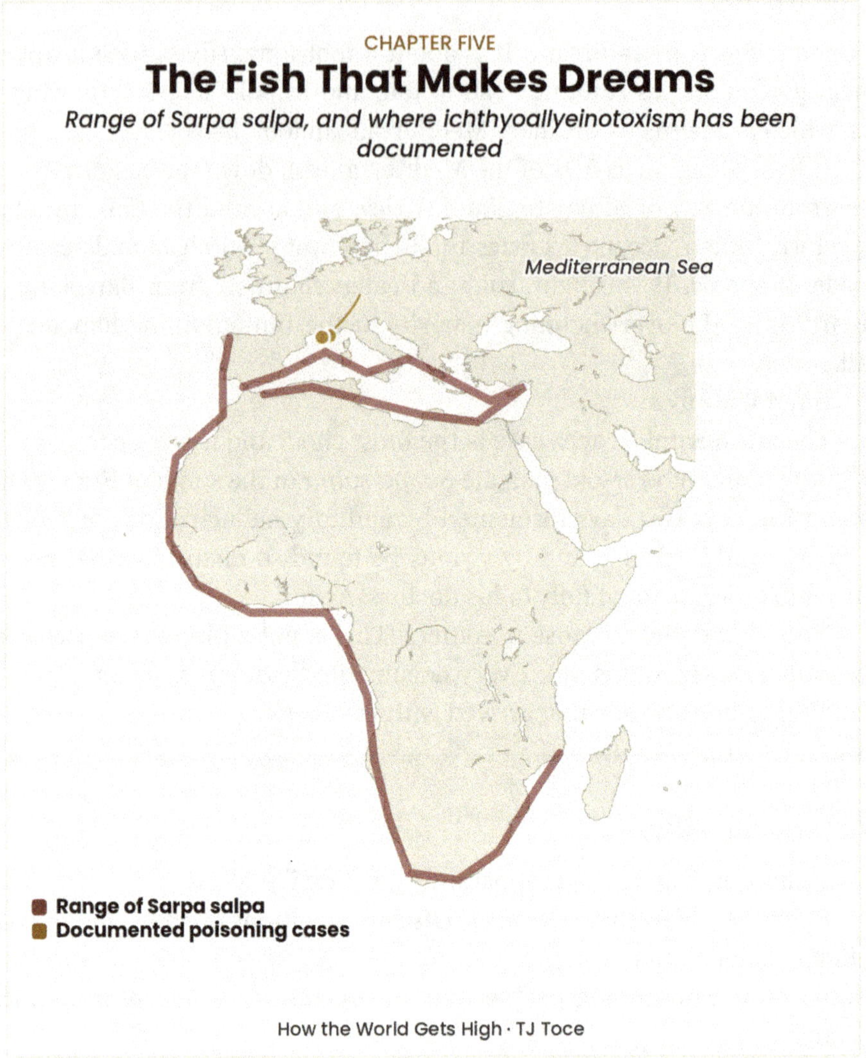


---

### THE DREAM FISH OF NORFOLK ISLAND

Our little Riviera fish is not alone.

Around Norfolk Island in the South Pacific, and elsewhere in the region, there is a sea chub of the genus *Kyphosus* known locally as the **dream fish**. Eating it is said to produce vivid and disturbing dreams and, in some accounts, waking hallucinations.

Sea mullet in parts of Australia have generated similar reports. So have a scattering of other species, including certain rabbitfish and surgeonfish, and, in a few accounts from the Indian Ocean, a species of goatfish.

There is a wonderful thread running through the folklore. In several traditions, these fish were reserved for elders, or for ceremonial occasions, or avoided entirely by fishing communities who knew perfectly well which species to throw back.

That is not superstition. That is a strategic food safety system, encoded in story, working for centuries without a laboratory anywhere in sight.

---

## THE TWIST: IT PROBABLY ISN'T THE FISH

And here is the turn, and it is a beautiful one, because it ties this chapter straight back to Chapter Three.

*Sarpa salpa* is a grazer. It spends its life cropping algae and seagrass off rocks in shallow coastal water.

The leading scientific explanation for ichthyoallyeinotoxism is that the fish is not producing anything at all. It is ACCUMULATING something. Specific species of algae or phytoplankton, present seasonally and regionally, are believed to contain compounds, possibly indole-based, that concentrate in the fish's flesh and organs.

This elegantly explains everything that is otherwise confusing about the phenomenon:

**1.** Why it is seasonal. Most reported cases cluster in the warmer months, when the relevant algae bloom.

**2.** Why it is regional. Fish from one stretch of coast cause it and fish from another never do.

**3.** Why it is so rare. The overwhelming majority of these fish, in the overwhelming majority of places, at the overwhelming majority of times, have grazed on nothing remarkable whatsoever.

**4.** Why the head is implicated. Traditional accounts consistently warn against eating the head and organs, which is exactly where such compounds would concentrate.

So the fish is a courier. It is carrying something a plant made, exactly the way a bee carries something a flower made, which is precisely where we are headed in the next chapter.

Are you seeing the pattern? It keeps being the plants. It is almost always the plants.

The kingdoms in this book are not separate rooms after all. They are connected by doorways, and the chemistry walks between them.

---

### FIELD NOTES: *SARPA SALPA*

**WHERE ON EARTH:** The entire Mediterranean Sea. The eastern Atlantic from the Bay of Biscay south to South Africa. Around the Cape into the western Indian Ocean.

**WHERE IT'S LEGAL:** Completely legal, everywhere. It is a food fish sold in ordinary markets throughout southern Europe and North Africa.

**HOW LONG WE'VE USED IT:** Reported since Roman antiquity, roughly two thousand years, with Pliny the Elder among the earliest sources.

**HOW IT'S OBTAINED:** Caught commercially and recreationally and sold in fish markets. Effects are unpredictable, seasonal, geographically specific and entirely accidental. This is the only entry in this book where the usual method of obtaining it is ordering lunch or dinner.

---

### THE STRAIGHT DOPE

Nobody should go looking for this.

Every documented case describes a frightening, disorienting episode lasting up to thirty-six hours, arriving without warning in someone who had no idea it was coming and no way to prepare. There is no dose control, no predictability, and no way to know in advance which fish carries what.

This is a marvelous piece of natural history, period. Facts first.

---

## THE POINT OF THE FISH

So why give a whole chapter to something so rare, unreliable and not worth pursuing?

Because it proves something the rest of this book keeps circling.

The chemistry is just OUT THERE. Woven through the food web, moving between kingdoms, showing up in a bream and a bee and a burrowing toad and a mushroom in a cow pasture in Wales.

We are not the authors of this story. We are latecomers who learned to read it.

That is the most humbling and the most wonderful thing in this entire book.

# CHAPTER SIX

## HONEY

*A flower's exotic chemistry, laundered through a bee*

> *Hast thou found honey? eat so much as is sufficient for thee, lest thou be filled therewith, and vomit it.*
>
> **PROVERBS 25:16 King James Version, 1611**

In the spring of 401 BCE, ten thousand Greek soldiers were having a very bad year.

They were mercenaries who had marched into Persia to help a prince seize a throne. The prince lost the battle and his life. The Greek generals were seized and executed soon after. And now the survivors were somewhere in eastern Anatolia, a thousand miles from home, in hostile country, with winter fast approaching.

One of the officers who took command was an Athenian named Xenophon, and because he later wrote the whole thing down in a book called the *Anabasis*, we have one of the greatest adventure stories in all of literature.

We also have the oldest detailed trip report in the Western world.

### THE SOLDIERS AND THE BEEHIVES

The army had fought its way north and reached the mountains above Trabzon, near the Black Sea coast, in the territory of a people called the Colchians.

They found villages. And in the villages, they found beehives.

The men, who had been marching and starving for months, did what starving soldiers do. They ate the honey.

Xenophon's account is precise and it is wonderful:

*All the soldiers who ate of the honeycombs lost their senses, and were seized with vomiting and purging, and none of them was able to stand upright. Those who had eaten only a little were like men very drunk, and those who had eaten much were like madmen, and some like dying persons.*

He continues. Great numbers lay on the ground as if after a defeat. Nobody died. And then:

*The next day, about the same hour, they came to their senses, and on the third or fourth day they got up as if from a drugging.*

Ten thousand men, flat on their backs in a Turkish mountain village, twenty-four centuries ago, taken out not by the Persian army but by breakfast.

## AND THEN SOMEBODY WEAPONIZED IT

Consider this. Three hundred and thirty-four years later, in 67 BCE, the same honey turned up in the same region as a weapon of war.

The Roman general Pompey was campaigning against Mithridates VI of Pontus, one of Rome's most persistent and creative enemies. Mithridates, who was a serious student of poisons and reportedly dosed himself with small quantities his entire life to build immunity, had local allies called the Heptakometes.

According to the historian Strabo, they set combs of the local honey out along the Roman line of march, in the open, as though abandoned. Of course, the Roman soldiers ate it.

And when they were incapacitated, staggering and vomiting and unable to hold a line, the Heptakometes came out of the hills and killed them all.

The first documented chemical weapon in military history was honey. How wild is that?

## WHAT IS ACTUALLY GOING ON

The culprit is a plant, as it usually is.

Certain rhododendrons, principally *Rhododendron ponticum* and *Rhododendron luteum*, produce compounds called **grayanotoxins**. These plants blanket the humid mountain slopes along the Black Sea coast of Turkey and Georgia, and they also grow across the Caucasus, in parts of Nepal, in Japan, and in a few scattered populations elsewhere.

When those rhododendrons are the dominant bloom, which happens in a narrow window each spring, bees foraging the area collect nectar loaded with grayanotoxin, and it ends up in the honey.

The Turkish name for it is **deli bal**. Mad honey.

Grayanotoxins bind to sodium channels in nerve and muscle cells and hold them open, which prevents the cells from resetting properly. The effects come on within thirty minutes to three hours: dizziness, tingling around the mouth and extremities, blurred vision, profuse sweating, nausea and vomiting, weakness, and a distinctly euphoric, floating, drunken sensation.

And critically, it slows the heart. Bradycardia and low blood pressure are the clinical hallmarks, and they are what usually put people in the hospital.

---

### THE HONEST VERDICT

This book made you a promise about honesty, so here it is.

**Mad honey is not really a psychedelic.**

There are no visions in the classical sense. There is no serotonin receptor involvement. There is no dissolution of self, no geometric imagery, no sense of encountering something larger. What there is, at moderate dose, is dizziness, a warm euphoric floating feeling, a slowed heart, and a fair chance of vomiting.

At higher doses it is a cardiac emergency.

So why is it in this book at all?

Because it is unquestionably one of the ways the world gets high, it has been for at least twenty-four hundred years, it is still sold openly by the spoonful in Turkish markets, and a book called *How the World Gets High* that left it out would be a poorer book. It earns its chapter on history alone.

But you deserve to know exactly what it is and why it is not worth your time. Facts first.

---

### THE STRAIGHT DOPE

The therapeutic window here is measured in teaspoons.

> Turkish folk practice uses roughly one teaspoon, typically on an empty stomach in the morning, and traditionally in the cooler months only. Turkish emergency departments see a steady stream of grayanotoxin poisoning cases every year, overwhelmingly in middle-aged and older men who took a tablespoon or three instead, generally in pursuit of its reputation as a folk remedy for blood pressure and sexual performance.
>
> Presentation is bradycardia, hypotension, and occasionally full cardiac block. It is treatable with atropine and fluids, and hospitalized patients almost always recover fully within a day or two.
>
> But the dose that makes you pleasantly dizzy and the dose that stops your heart from doing its job properly are uncomfortably close together, and the concentration varies wildly from jar to jar with no way to test it at home.

---

### THE HONEY HUNTERS OF NEPAL

Now let's go somewhere truly spectacular.

In the foothills of the Annapurna range in central Nepal, at altitudes between five and twelve thousand feet, lives *Apis dorsata laboriosa*, the Himalayan giant honey bee. It is the largest honey bee on Earth, up to an inch and a quarter long, and it does not live in hives.

It builds enormous single combs, some of them five feet across, hanging from the underside of overhanging cliff faces, hundreds of feet above the valley floor. The bees choose these sites precisely because nothing can reach them.

Almost nothing.

Twice a year, the Gurung people of Nepal harvest these combs in a practice that is, without exaggeration, one of the most extraordinary things any human does anywhere.

Here's how it goes. A rope ladder, handmade from bamboo and hemp, is anchored at the clifftop and dropped down the rock face. A harvester climbs down onto it, sometimes two hundred feet or more above the rocks below, wearing minimal protection or none at all. Below, a team builds smoky fires to drive off the bees. The harvester cuts the comb free with a long bamboo pole and guides it into a basket lowered on a second rope.

He does this while hanging on a swinging ladder, in smoke, surrounded by the largest honey bees in the world, which are furious. I've seen videos of this and it is beyond scary!

And the honey he brings up, in the spring harvest when the rhododendron is in bloom on those slopes, is red-tinted mad honey.

The Gurung call it by different names in different valleys. It is used medicinally, traded, and increasingly sold to a global market. The harvest is accompanied by ceremony and offerings, because everyone involved understands exactly what is being risked.

Is that not one of the most unbelievable things you have ever heard? People climbing down a cliff on a bamboo ladder, into a cloud of giant bees, for honey that makes you dizzy.

Humans are remarkable. We really are.

---

## THE TURKISH BLACK SEA COAST TODAY

Back in Turkey, the tradition never stopped and never went underground, because it never needed to.

The provinces along the eastern Black Sea coast, Rize, Artvin, Trabzon, Giresun, are steep, wet, green and covered in rhododendron. Beekeeping there is ancient. And deli bal is produced deliberately by beekeepers who move their hives up to the rhododendron slopes during the bloom.

It is sold in local markets, at roadside stands, and in shops in Istanbul, and it commands a substantial premium over ordinary honey. It is taken as a folk remedy for hypertension, for digestive complaints, for arthritis, and for sexual performance, and the customer base skews strongly toward men over fifty.

**And it is entirely legal in the United States.** Grayanotoxin is not a controlled substance anywhere. Mad honey is imported, sold on eBay and Etsy and dedicated websites, and shipped to American customers right now, today, in jars labeled exactly what they are.

A substance that took out ten thousand Greeks and killed a thousand Romans, available with free shipping. What a time to be alive.

---

## FIELD NOTES: MAD HONEY

**WHERE ON EARTH:** Produced wherever grayanotoxin-bearing rhododendrons dominate the spring bloom. Principally the eastern Black Sea coast of Turkey and Georgia, and the Annapurna foothills of Nepal. Smaller occurrences in the Caucasus, Japan, Korea, parts of Brazil, and occasionally in the Pacific Northwest of the United States from *Rhododendron macrophyllum* and mountain laurel.

**WHERE IT'S LEGAL:** Everywhere. Grayanotoxin is not a controlled substance in any jurisdiction. Legal to produce, sell, import and possess in the United States, Europe, Turkey and Nepal.

**HOW LONG WE'VE USED IT:** Documented since 401 BCE via Xenophon, and used as a weapon by 67 BCE. Continuous folk use in Turkey and Nepal ever since. Twenty-four hundred years and counting.

**HOW IT'S OBTAINED:** Purchased in Turkish markets, harvested from Himalayan cliff faces by Gurung honey hunters, or ordered online and delivered to your door. Concentration is entirely unregulated and varies enormously between batches.

## WHAT THE HONEY TEACHES

Here is why this chapter matters more than its subject might suggest.

A flower was the creator. The rhododendron made the grayanotoxin, almost certainly to discourage insects and animals from eating it. The bee simply harvested nectar and did what bees do. The honey is a plant's chemical defense system, collected, concentrated and repackaged by an entirely different kingdom of life, and then handed to a third one.

Exactly like the fish in Chapter Five, carrying what the algae made.

Do you see it? The chemistry does not respect our categories. It is always on the move. Flower to bee to jar to the curious adventurer seeking a new experience.

CHAPTER SIX

## The Mad Honey Belt

*The two regions on Earth that produce grayanotoxin honey at scale*

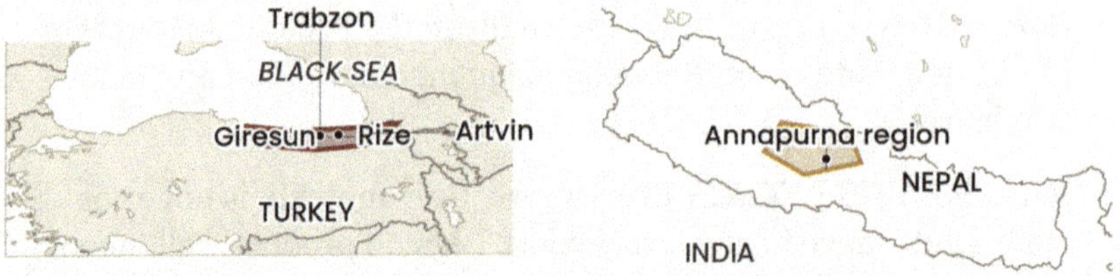


How the World Gets High · TJ Toce

CHAPTER SIX

## From Flower to Spoon

*How a plant's chemical defense ends up in a jar*

**RHODODENDRON**
Produces grayanotoxin in leaf and nectar
*to deter insects and grazing animals*

›

**NECTAR**
Toxin present in the nectar of R. ponticum
*only during the spring bloom*

›

**THE BEE**
Forages normally. Unaffected herself
*carries it back to the comb*

**THE COMB**
Honey now carries the toxin
*concentration varies wildly, batch to batch*

›

**THE SPOON**
One teaspoon: folklore. Three: bradycardia
*sold openly in Turkey and online*

*Everything in this chapter came from a flower.*

*The rhododendron made the toxin. The bee, which has no interest in any of this, simply moved it.*

How the World Gets High · TJ Toce

# CHAPTER SEVEN

## THE HUMAN BODY

*The hardware was installed at the factory*

> *The mind, in discovering truth, acts in the same manner as it acts through the eye in discovering objects; when once any object has been seen, it is impossible to put the mind back to the same condition it was in before it saw it.*
>
> **THOMAS PAINE** *Rights of Man, 1791*

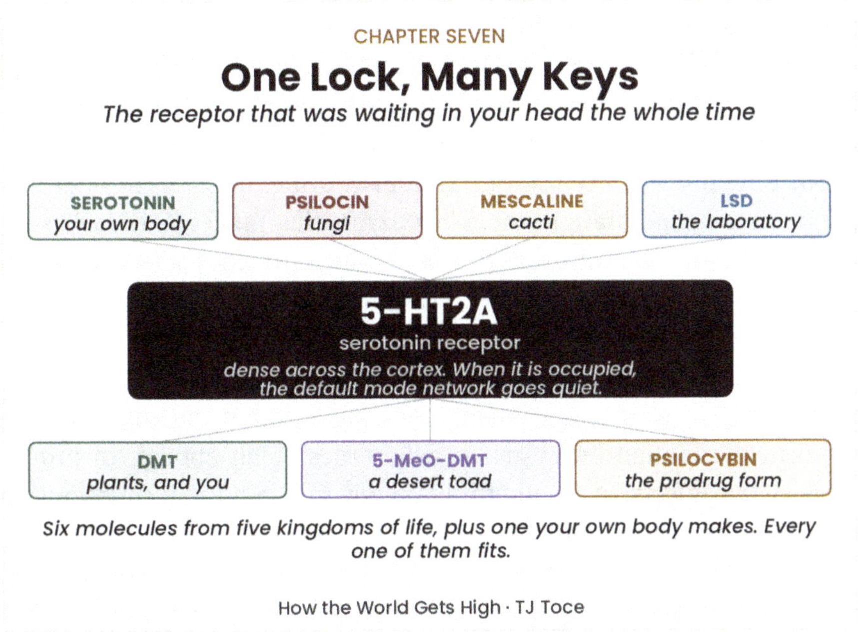


Here is where the whole book takes a fascinating turn.

We have traveled through six chapters looking OUTWARD. A mushroom in a Welsh pasture. A vine in the Amazon. A cactus in the Chihuahuan Desert. A toad in a Sonoran burrow. A bream off Saint-Tropez. A jar of honey from the Black Sea coast and Nepal.

All of it external. All of it something you find, harvest, brew, dry or buy.

Now put all of that down for a moment and consider something much closer to hand.

Your own body manufactures at least one of these compounds. Your own brain can enter these states with nothing added at all. And humans have known how to trigger them, deliberately and reliably, without any plant or fungus whatsoever, for as long as we have known anything.

The hardware was installed at the factory but what about the owner's manual? Follow me.

---

## THE MOLECULE INSIDE

In 1965, researchers detected DMT in human blood.

That is worth stopping on. DMT, the compound in the ayahuasca leaf, the compound in the toad, one of the most powerful psychedelics known to science, is manufactured by the human body.

Since then it has been found in human urine, in cerebrospinal fluid, and in mammalian brain tissue. The enzyme that makes it, INMT, is present in human lung, brain and other tissues. In 2019 a research team demonstrated that the rat brain actively synthesizes DMT and that levels can rise under certain conditions.

So the fact is established. Our bodies make it.

What we do NOT know is what it is for, whether the concentrations found in the human brain are anywhere near high enough to produce psychoactive effects, or whether it plays any role in normal consciousness at all. But oh the possibilities.

The quantities detected are extremely small. Some researchers argue it is a metabolic byproduct with no function. Others argue it is highly localized and released in bursts we have not yet learned to measure. The question is genuinely open.

What we can say for certain is this: the human body contains the enzymatic machinery to produce a classical psychedelic, and it does so.

Whatever that means, it is not nothing.

---

## STRASSMAN AND THE PINEAL

You cannot discuss this without discussing Rick Strassman, and it is a story worth telling.

In 1990, Strassman, a psychiatrist at the University of New Mexico, an absolutely beautiful school, obtained approval to administer DMT to human volunteers. It was the first new human psychedelic research in the United States in over two decades, and getting it approved took him two years of paperwork and considerable tenacity.

Between 1990 and 1995 he ran roughly four hundred doses through sixty volunteers.

What his subjects reported was consistent and, for a clinical setting, startling. Extremely rapid onset. A sense of traveling somewhere. And, in a large fraction of sessions, the conviction of encountering intelligent "beings" who appeared to be aware of and interested in the volunteer. Tapping into another dimension perhaps?

Strassman was a rigorous scientist and he did not know what to do with this. His 2000 book *DMT: The Spirit Molecule* laid out the data alongside a speculative hypothesis: that the pineal gland, a small structure near the center of the brain, produces DMT and releases it at birth, at death, and during profound spiritual experience. You may know it as the third eye.

That hypothesis has been enormously influential in popular culture but is not yet supported by evidence.

The pineal gland's known job is producing melatonin. Trace DMT has been found there, but also in other tissues, and there is no demonstration of the surge Strassman proposed. He himself has been careful and forthright about this, repeatedly stating that it was a hypothesis offered for testing, not a finding.

So here is a beautiful story that is probably a myth. But notice what survives once you strip the speculation away, because it is still remarkable: a psychiatrist gave sixty people a compound their own bodies also make, and a large share of them came back describing the same extraordinary thing.

## WHAT HAPPENS AT THE EDGE

Roughly one in ten people who survive cardiac arrest report a near-death experience, and the reports are startlingly consistent across cultures, religions and centuries.

The elements recur: a sense of leaving the body and observing it from outside. Movement through a dark space, often described as a tunnel. A brilliant light. Encounters with deceased relatives or with a being of overwhelming benevolence. A rapid review of one's life. A boundary that cannot be crossed. And a profound reluctance to return.

Afterward, survivors very frequently report reduced fear of death, altered priorities, and increased sense of meaning, effects that persist for years.

What causes this?

Several explanations compete, and they are not mutually exclusive. Oxygen deprivation produces tunnel vision and euphoria. Surges of endorphins and endogenous opioids occur under extreme stress. And yes, there is the DMT hypothesis, still unproven.

The most interesting research finding, in my opinion, is this. In 2018 a study directly compared the written accounts of people who had near-death experiences with the accounts of people given high-dose DMT.

The similarity was striking enough that the researchers could not reliably tell them apart.

What that means is genuinely unknown. But something in us knows how to do this, and it seems to know how to do it precisely when everything else is going wrong.

---

## THE BLISS MOLECULE

Let's lighten the mood, because your body has been keeping other secrets and this one is lovely.

For decades, the pleasant floating feeling that arrives after sustained exercise was attributed to endorphins. The word entered popular language as "endorphin rush" and everybody accepted it.

Then researchers noticed a problem. Endorphins are large molecules. They do not readily cross the blood-brain barrier. Whatever was producing that specific quality of euphoria, calm and mild alteration of perception, endorphins were most likely not the answer.

The answer turned out to be **anandamide**, an endocannabinoid your body produces that binds to the same receptors as THC.

And here is the part I love. It was isolated in 1992 by a team including Raphael Mechoulam, and they named it from the Sanskrit word *ānanda*.

Bliss.

Your body makes a bliss molecule. It releases it when you run, when you laugh, when you are with people you love, and, interestingly, when you eat dark chocolate. Bliss!

---

### THE BREATH

Now we get practical, because everything from here on is something a human being can do, for free, today, legally, anywhere on Earth.

Change your breathing and you change your consciousness. This is not mysticism. It is physiology, and it works every single time. Yogis have known this for millennia.

**Holotropic Breathwork** was developed in the 1970s by the Czech psychiatrist Stanislav Grof and his wife Christina. Grof had been one of the most prolific LSD researchers in the world, running clinical sessions in Prague and then in Maryland, and when LSD was scheduled in 1970 his entire research program evaporated.

So he went looking for another door.

He found it in accelerated breathing. Participants breathe deeply and rapidly, for an extended period, typically two to three hours, lying down, in darkness, with loud evocative music, and with a sitter present.

What people report is not a mild relaxation. It is the full territory: vivid imagery, powerful emotional release, autobiographical memory, and experiences Grof classified as perinatal and transpersonal, which is to say the same country the psychedelics visit.

Mechanistically, rapid breathing drops carbon dioxide levels, which changes blood pH and constricts cerebral blood vessels. The precise route from there to the reported experiences is not fully mapped, but real, very, very real.

**The Wim Hof Method** uses cycles of controlled hyperventilation followed by breath retention and cold exposure, and has been studied for its effects on the autonomic nervous system and immune response, with some genuinely surprising results.

Yogic **pranayama** traditions have been doing versions of this deliberately for well over two thousand years, and they built extremely detailed maps of the territory long before anyone measured a blood gas.

One caution and then we move on. This is powerful. It requires practice, and professional supervision is strongly recommended for beginners. Facts first.

---

### THE ONE YOU DO EVERY NIGHT

Consider this, and then consider that it happens to you for free, nightly, whether you like it or not.

Four or five times a night, you enter REM sleep. And in REM sleep, your brain does the following:

**1.** It generates a complete, immersive, three-dimensional world with no external sensory input whatsoever.

**2.** It populates that world with people who speak, act and surprise you.

**3.** It suspends your critical faculty so completely that you accept flying, or being someone else, or a dead relative sitting at the kitchen table, without a flicker of doubt.

**4.** It paralyzes your body so you do not act any of it out.

**5.** It erases nearly all of it within minutes of waking.

If a substance did that to you, it would be Schedule I by Friday.

You have hallucinated vividly every single night of your entire life. The machinery is not exotic and it is not borrowed. It is standard equipment, and it runs on a nightly schedule.

Lucid dreaming, in which the sleeper becomes aware they are dreaming and can direct the experience, is a real and reproducible phenomenon that has been confirmed in the laboratory through pre-agreed eye movement signals from sleeping subjects. People have trained themselves into it for centuries. Tibetan Buddhist dream yoga is built entirely around it. Beyond fascinating!

Your brain has a full studio in the basement, and it runs a show every night whether you come down or not.

---

## RHYTHM, AND WHY THE DRUM

Walk into a traditional healing ceremony almost anywhere on Earth and you will find a drum. Rock and roll is a latecomer to an ancient form of music.

Siberian shamanic practice. West African ceremony. Native American ritual. Sami tradition in northern Scandinavia. Aboriginal Australian practice. These traditions had no contact with each other, and they all arrived at the same technology.

The tempo tends to land between roughly four and eight beats per second, which corresponds to the brain's theta frequency band, the range associated with drowsiness, deep meditation and dreaming.

Laboratory studies of monotonous drumming at these tempos have found measurable changes in brainwave activity and consistent reports of altered states in listeners, including imagery and a sense of journeying.

Rhythm is not the only route. There is Sufi whirling, the Mevlevi practice founded in the tradition of Rumi in thirteenth-century Konya, in which sustained rotation produces a state practitioners describe as annihilation of the self in the divine. There is the trance dance of the San people of the Kalahari, which may be the oldest continuously practiced healing ceremony on Earth, in which dancers move around a fire for hours until healers enter a state called *kia*.

Different continents. Different centuries. No shared contact. All of them found a rhythm and used it as a door to an expanded mind.

Are you seeing the pattern? It really does keep happening.

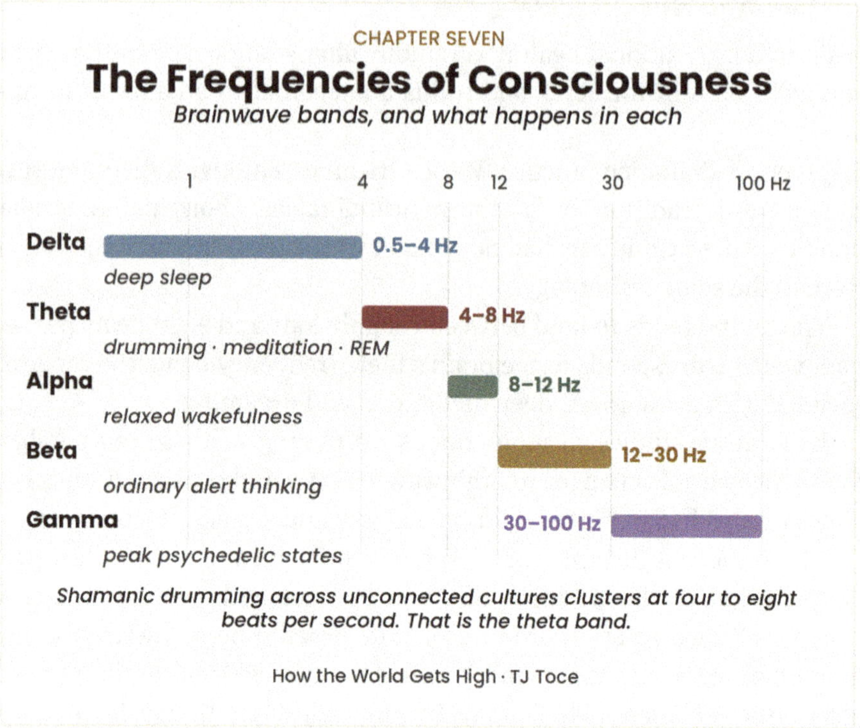


## SUBTRACTION

Everything above adds something: breath, rhythm, motion. The last family of techniques works by taking things away.

**Sensory deprivation.** In 1954 the neuroscientist John Lilly built the first isolation tank, floating subjects in body-temperature salt water in total darkness and silence to find out what a brain does with no input at all. What it does is generate its own. Modern float tanks are now found in most major cities and are entirely legal.

**Fasting.** Nearly every major religious tradition on Earth incorporates it, and prolonged fasting reliably produces altered mood, heightened perception and, past a certain point, visionary experience. Every tradition that uses it says the same thing about why.

**Sleep deprivation.** Vigils, all-night ceremonies, and the deliberate holding of wakefulness appear across an enormous range of traditions.

Extended sleeplessness produces perceptual disturbance in essentially everyone. Not very pleasant and not recommended.

**The vision quest.** Practiced across many Native American nations, though the details differ enormously and the term flattens a great deal of variety. Typically it involves solitude, exposure, fasting, and days without food or company in a natural place, undertaken with preparation and community support before and after.

**Pilgrimage, extreme exertion, and fever.** Every one of them a known route into altered perception. Every one of them used deliberately by somebody, somewhere, for a very long time.

---

## WHAT THIS CHAPTER IS REALLY SAYING

It's time once again to throw in a little context for clarity.

The rest of this book is about substances, and substances come tangled in law, in supply, in risk, in cost, in ceremony and in an enormous amount of controversy.

This chapter is about none of that.

Breath is free. Rhythm is free. Fasting is free. Dreaming happens whether you consent to it or not. Every technique described here is available to every human being alive, in every country on Earth, legally, right now, and has been since time began.

Which raises the question this chapter exists to raise. The big picture! If the machinery is standard equipment, and the doors are already in the walls, then what exactly are the plants and the fungi and the toad doing?

The most compelling answer researchers have offered is that they are not installing anything new. They are turning a key that was already there, in a lock that was already there, that your own body already knows how to turn.

The mushroom is not giving you something foreign. It is giving you something familiar, at a volume you cannot ignore.

Can you see the intention in that? Every kingdom of life we have visited in this book, arriving independently at a molecule shaped to fit a receptor that was waiting in your head the whole time.

You were designed and built for this. What an exhilarating and most extraordinary thought!

---

# CHAPTER EIGHT

## SYNTHETIC DRUGS

*What happened when we started building them ourselves*

> *All things are poison, and nothing is without poison; the dose alone makes a thing not a poison.*
>
> **PARACELSUS Septem Defensiones, 1538**

---

CHAPTER EIGHT

**Albert Hofmann**

*Lugano, October 1993*

Photographed at the conference marking fifty years since Bicycle Day.
*He called LSD his problem child, and lived to 102.*

*Photograph by Philip H. Bailey, CC BY-SA 2.5*

How the World Gets High · TJ Toce

---

For roughly ten thousand years, humans took what the planet, Gaia, offered. A mushroom in a pasture. A vine in a forest. A cactus in a desert. You went and you found it, and if it was not there, you went without.

Fast forward to the twentieth century, and then we started making our own.

That changed everything. Not just the potency and the availability, though it changed both dramatically. It changed the RELATIONSHIP. A mushroom has a season. A cactus takes thirty years. A laboratory has a Tuesday.

This chapter is about what we built, why, and what it cost us.

---

### LSD: THE PROBLEM CHILD

We have met Albert Hofmann twice already. Now let's give him the full chapter he has earned.

Hofmann was born in Baden, Switzerland in 1906, and he joined Sandoz Laboratories in Basel in 1929. His assignment was ergot, the rye fungus from Chapter Two, which had been feared for centuries and was now understood to contain alkaloids of genuine medical value. Midwives had used crude ergot preparations to control postpartum bleeding for generations. Sandoz wanted to isolate what worked and leave behind what killed.

Hofmann synthesized a long series of lysergic acid derivatives. In November 1938 he produced the twenty-fifth, lysergic acid diethylamide, and labeled it LSD-25.

The pharmacology department tested it. The animals became restless during the anesthesia phase, which was noted as mildly interesting. There was no other effect worth pursuing. The compound was shelved.

And that should have been the end of it.

Five years later, in April 1943, Hofmann did something no chemist is really supposed to do. He went back to a discarded compound for no reason he could articulate.

He wrote later: *A peculiar presentiment, the feeling that this substance could possess properties other than those established in the first investigations, induced me to produce it once again.*

A presentiment. Very odd word, yet that is the word he used, in a scientific memoir, about the most consequential resynthesis in the history of chemistry. A presentiment is a sense of something before it arrives, and usually something bad. From the Latin, to feel beforehand.

On April 16 he was working with the compound and became strangely restless and dizzy. He went home, lay down, and experienced what he described as an uninterrupted stream of fantastic pictures with intense, kaleidoscopic play of colors.

He suspected the compound but could not believe it, because no substance known to science was active in the trace amounts he could possibly have absorbed through his fingertips. How could this be?

So on Monday April 19, he tested it properly. He recorded in his lab journal that at 4:20 in the afternoon he took 250 micrograms dissolved in water, a dose he considered extremely conservative.

It was, in fact, a very strong dose. Nothing in the pharmacology of the era had prepared anyone for a substance active at these levels.

Forty minutes in, his journal entry breaks off mid-sentence. He asked his laboratory assistant to help him home. Wartime restrictions meant no car was available.

So they rode bicycles through Basel, and Hofmann later described the four-mile ride as a journey in which the world warped and streamed and his neighbor turned into a malevolent witch, and he was convinced he had poisoned himself and was dying.

By evening it had turned. He described the fear lifting and being replaced by wonder, and he wrote of lying in bed watching fantastic images surge in on him, in circles and spirals, exploding in colored fountains.

The next morning he woke feeling, in his words, refreshed, with a clear head, and walked into the garden after a spring rain shower where everything glowed in fresh light.

April 19 is now commemorated worldwide as **Bicycle Day**. It remains the only holiday on Earth founded on a chemist's ride home from work.

## WHAT SANDOZ DID NEXT

Here is the part that always surprises people.

Sandoz brought LSD to market in 1947 under the trade name **Delysid**, and their strategy was to send it to psychiatric researchers around the world, essentially free of charge, with a request that they report back what they found.

Think about that business model for a moment.

Over the following fifteen years, more than a thousand scientific papers were published, involving tens of thousands of patients. It was studied for alcoholism with results that looked genuinely promising. It was studied for depression, for obsessional neurosis, for anxiety in terminal cancer patients. It was used as a tool to help therapists understand psychosis by temporarily experiencing something adjacent to it.

Cary Grant took it roughly a hundred times in supervised therapy and gave interviews to *Look* magazine saying it had made him a happier man and a better one.

For about twenty years this was simply ordinary medicine. Boring. Peer reviewed. Government funded.

Then it left the building.

**MK-Ultra.** Beginning in 1953, the CIA ran a program to investigate LSD as a tool for interrogation and behavioral control. It involved administering the drug to people without their knowledge or consent, including prisoners, psychiatric patients, and members of the public. One participant, an army scientist named Frank Olson, died in a fall from a hotel window in 1953 in circumstances his family has disputed ever since. Most of the records were deliberately destroyed in 1973.

It is a genuinely appalling chapter and it deserves to be spotlighted here.

**Harvard.** More publicity. Timothy Leary and Richard Alpert began running psilocybin research at Harvard in 1960. The science started reasonably and drifted. Leary proved to be an extraordinary communicator and a poor experimentalist, and his gift for publicity did enormous damage to colleagues who were trying to do careful work. Both men were dismissed in 1963.

**And then the culture war.** By 1966 LSD was on magazine covers and in congressional hearings. Sandoz stopped distributing it. States began banning it.

On October 27, 1970, the Controlled Substances Act became law. LSD, psilocybin, mescaline, DMT and peyote went into Schedule I: high potential for abuse, no accepted medical use, no accepted safety even under supervision. What was behind all this governmental fear?

Hofmann lived until 2008, dying at the age of 102. He never stopped arguing that the compound had been misunderstood, and he titled his memoir *LSD: My Problem Child.*

He also lived long enough to see the research restart. In 2007, at the age of 101, he learned that a new clinical trial of LSD had been approved in Switzerland.

He said he was very glad. What a life. What a long wait.

The future of **LSD-25** is now in the hands of medical science and psychiatric treatment. After decades of suffocating legal restrictions that halted all studies, LSD-25 is undergoing a major clinical renaissance all over the world.

---

## FIELD NOTES: LSD

**WHERE ON EARTH:** Entirely synthetic, though derived from ergot alkaloids. Manufactured in clandestine laboratories worldwide, and legally in a small number of licensed research facilities.

**WHERE IT'S LEGAL:** Schedule I in the United States and controlled in nearly every country. Legal for approved clinical research in Switzerland, the United States, the United Kingdom, Canada and elsewhere. Switzerland operates a limited compassionate-use program permitting a small number of physicians to administer it.

**HOW LONG WE'VE USED IT:** Synthesized 1938. First deliberate human dose April 19, 1943. Eighty-three years.

**HOW IT'S OBTAINED:** Illicit manufacture and distribution, typically on blotter paper. Enrollment in a clinical trial. Nothing else.

---

## MDMA: THE ONE THAT WAITED SIXTY YEARS

MDMA has the strangest history of anything in this book, because it was invented, forgotten, reinvented, embraced, banned, and then walked back to the door of the FDA.

**1912.** The German pharmaceutical company Merck synthesized it as an intermediate step while working toward a compound to control bleeding. It was patented in 1914. Nobody tested it on a human being. It sat in the archive for over half a century.

**1976.** Alexander Shulgin, a Berkeley chemist we will meet properly in a moment, resynthesized it after a student mentioned it to him, and tried it himself.

What he found was not a classical psychedelic. There were no significant visual effects. What there was, he reported, was an extraordinary opening: a collapse of anxiety and defensiveness, an enormous increase in emotional warmth, and the ability to look directly inside oneself at painful memories and start analyzing them without flinching.

Shulgin recognized immediately what that might be worth, and he did something unusual. He did not publish widely. He introduced it, quietly and personally, to a psychologist named Leo Zeff.

Zeff was on the edge of retirement. He quickly unretired. Over the following years he introduced MDMA to over four thousand therapists, and it spread through the psychotherapy community by word of mouth, entirely legally, because nobody had scheduled it.

Therapists initially called it **Adam**, because of the state of innocence patients described.

For roughly a decade, thousands of therapists used it in practice with results they described as remarkable, particularly for trauma and for couples work.

**Then it got a different name, or some would say its true name.** By the early 1980s it had escaped into nightclubs in Dallas and on the coasts, where it was called **Ecstasy** and sold openly, briefly, even on bar tabs. Go Cowboys!

Not surprisingly, in 1985 the DEA moved in and placed it in Schedule I on an emergency basis. Therapists and researchers fought it. An administrative law judge recommended Schedule III, which would have preserved medical use. The DEA overruled him. Why?

**1986.** Rick Doblin founded the Multidisciplinary Association for Psychedelic Studies, MAPS, with the explicit long-term goal of making MDMA a legal prescription medicine. He was twenty-nine years old. He has been working on it to this very day.

MAPS ran the trials, phase by phase, over decades. Phase 3 results published in 2021 and 2023 showed substantial reductions in PTSD symptoms, with a large fraction of participants no longer meeting the diagnostic criteria at all.

In June 2024, an FDA advisory committee voted against recommending approval, citing concerns about trial blinding, therapist conduct in some sessions, and data integrity. In August 2024 the FDA issued a complete response letter requesting an additional trial.

The story did not end there. In April 2026 the FDA issued a National Priority Voucher to Transcend Therapeutics for **methylone**, an MDMA-adjacent compound being developed for PTSD, putting it on a dramatically accelerated review path.

One hundred and fourteen years from a German patent nobody read to a priority review voucher. Patience is a chemical property, apparently.

CHAPTER EIGHT

## Alexander Shulgin

*Austin, Texas, 2010*

He designed and tested more than two hundred novel compounds, *and never asked anyone to take one he had not taken first.*

*Photograph by Charlie Llewellin, CC BY-SA 2.0*

How the World Gets High · TJ Toce

---

### FIELD NOTES: MDMA

**WHERE ON EARTH:** Entirely synthetic. Precursors historically derived from safrole, found in sassafras and related plants, which is why sassafras oil is a watched chemical.

**WHERE IT'S LEGAL:** Schedule I in the United States. Legal for approved clinical research in numerous countries. Australia authorized prescription use by approved psychiatrists for PTSD beginning July 2023, becoming the first country to do so.

**HOW LONG WE'VE USED IT:** Synthesized 1912. First documented human use 1976. Therapeutic use 1977 to 1985. Fifty years of human experience.

**HOW IT'S OBTAINED:** Illicit manufacture. Clinical trial enrollment. Prescription in Australia under the authorized prescriber scheme.

---

## KETAMINE: THE ONE YOU CAN ACTUALLY GET

Now for the outlier, and the only compound in this chapter with a legal clinic somewhere near you right now.

Ketamine was synthesized in 1962 by Calvin Stevens at Parke-Davis, as a replacement for PCP, which worked as an anesthetic but left patients agitated and disturbed for hours afterward. Ketamine was shorter-acting and far more manageable.

It was approved by the FDA in 1970 and used extensively as a battlefield anesthetic in Vietnam, because it has an enormous safety margin and does not suppress breathing the way most anesthetics do. It remains on the World Health Organization's List of Essential Medicines for exactly that reason. In much of the world it is the anesthetic available when nothing else is.

Anesthesiologists noticed early that patients emerging from ketamine reported vivid dream states, out-of-body sensations, and encounters they struggled to explain. This was categorized as an unwanted side effect and called emergence phenomena.

**Then, in 2000, a small study at Yale found something nobody expected.** A single low dose of ketamine produced rapid, substantial relief from severe depression, in some patients within hours.

Understand why that was startling. Conventional antidepressants take four to six weeks to kick in and fail entirely for roughly a third of patients. Here was something that worked in an afternoon on almost everyone.

The finding held up. Ketamine works through the glutamate system rather than serotonin, and appears to promote rapid regrowth of synaptic connections in brain regions damaged by chronic stress.

In March 2019 the FDA approved **esketamine**, a nasal spray, for treatment-resistant depression. And because generic ketamine had been an approved anesthetic for fifty years, physicians could prescribe it off-label, which is exactly what happened.

Ketamine clinics now operate in every major American city. Telehealth providers prescribe at-home protocols. High profile users have turned this into a real industry with some very real oversight questions.

Its legal position is a genuine historical accident. It was approved as an anesthetic in 1970, months before the Controlled Substances Act took effect the following May, and so it never went into Schedule I. It sits in Schedule III.

The most available psychedelic-adjacent medicine in America is legal because of a timing coincidence. Isn't that something?

---

### FIELD NOTES: KETAMINE

**WHERE ON EARTH:** Entirely synthetic, manufactured pharmaceutically worldwide.

**WHERE IT'S LEGAL:** Schedule III in the United States, legally prescribable. Esketamine is FDA-approved for treatment-resistant depression. Widely legal as a prescription anesthetic globally and listed by the WHO as an essential medicine. Non-medical possession remains illegal.

**HOW LONG WE'VE USED IT:** Synthesized 1962. FDA-approved 1970. Recognized as an antidepressant since 2000.

**HOW IT'S OBTAINED:** Prescription. Infusion clinics, psychiatric practices and telehealth providers throughout the United States. This is the only entry in this chapter obtainable by making an appointment.

---

### SHULGIN, AND THE MAN WHO TESTED THEM ALL

Alexander Shulgin deserves his own section and special recognition, because there has never been anyone quite like him and there very likely never will be again.

He was a biochemist at Dow Chemical who developed a highly profitable biodegradable pesticide. Dow was grateful and gave him unusual freedom. He used it to research psychedelics, which Dow tolerated until it didn't, and he left in 1966 to work independently from a laboratory in a shed behind his house in Lafayette, California.

And here is the extraordinary part: he held a DEA Schedule I research license. He operated legally, for decades, synthesizing controlled substances in a backyard shed, and he consulted for the DEA, taught at their training academy, and testified as an expert witness for them.

Over his career he designed, synthesized and tested more than two hundred novel psychoactive compounds.

Tested on whom? On himself, and on his wife Ann, and on a small circle of trusted friends who met regularly and rated everything on a scale Shulgin created.

He started at doses so low they could not possibly have an effect and worked up in careful increments, documenting everything.

In 1991 he and Ann published ***PiHKAL: A Chemical Love Story***. The title stands for Phenethylamines I Have Known And Loved. It is two books in one: a lightly fictionalized memoir of their life together, and a technical section giving the full synthesis, dosage and effects for 179 compounds.

*TiHKAL*, covering tryptamines, followed in 1997 with 55 more.

He published complete synthesis instructions for controlled substances. Openly. Under his own name. While holding a DEA license.

The DEA raided him in 1994, fined him, and he surrendered the license. He kept working until the day he died in 2014 at the age of 88.

His legacy is genuinely double-edged and both edges are sharp. He gave the world the 2C family of compounds, several of which have real research interest. He also handed a recipe book to an industry that would later mass-produce these drugs with little or no safety data at all.

He was asked about this. His answer was consistent: information should be free, and the responsibility for what people do with it belongs to them.

You may agree or disagree. He never wavered, and he never asked anyone to take a compound he had not taken first. Definitely not your standard approach.

## SIDEBAR: HOW TO READ A TRIP REPORT

# HOW TO READ A TRIP REPORT

You have already read four of them in this book.

Xenophon's ten thousand Greeks flat on their backs in a Turkish mountain village in 401 BCE. That is a trip report. Albert Hofmann's laboratory journal breaking off mid-sentence on April 19, 1943. Trip report. Aldous Huxley looking at three flowers in a glass vase in 1953 and writing sixty pages about it. Trip report. R. Gordon Wasson sitting in a dark room in Oaxaca in 1955 listening to María Sabina sing. Trip report, and one that went out to five and a half million readers.

Every one of them is somebody trying to write down something that resists being written down. And every one of them shaped how the world understood this subject for generations afterward.

The tradition never stopped. It just moved online.

## THE ARCHIVE NOBODY PLANNED

Here is one of the genuinely wonderful accidents of the prohibition era.

When formal research shut down in 1970, the questions did not. People kept having these experiences, and people kept writing them down. Beginning in the 1990s, those accounts started collecting online, principally on Erowid, which has been operating since 1995 and now holds tens of thousands of firsthand reports. Add Reddit, Bluelight, DMT-Nexus and a dozen others and the total runs into the hundreds of thousands.

Think about what that represents. During twenty-five years when the scientific establishment was not permitted to ask the question, ordinary people built the largest library of human psychedelic experience that has ever existed.

And researchers everywhere are using it. Published papers have mined these archives for rare interactions and adverse effects that no clinical trial had the numbers to catch. When a new compound appears on the market, the first useful safety information about it almost always shows up in a trip report before it shows up in a journal.

That archive is a gift. It is also a minefield, and knowing how to walk through it is a real skill.

## SHULGIN SAW THE PROBLEM FIRST

Alexander Shulgin tested more than two hundred novel compounds, and he understood immediately that "it was intense" and "it was really something" are worthless if you are trying to compare one molecule to another.

So he built a scale.

**MINUS** – no effect at all. **PLUS/MINUS** – something may be happening, or it may be your imagination. **PLUS ONE** – definitely active, but no sensory change. **PLUS TWO** – clear effects on perception and thought, still controllable. **PLUS THREE** – fully engaged, impossible to do anything else. **PLUS FOUR** – a rare and separate category Shulgin reserved for the genuinely transcendent, and he was clear that it could not be reached by increasing the dose.

He and Ann rated every single compound this way, in the same conditions, with the same small group, over decades.

That is what turns a story into data. And it is the gold standard to apply to all.

## SEVEN THINGS TO ASK OF ANY REPORT

**1. Did they actually know what they took?** Almost nobody tests. A report titled "LSD" may describe something that was never LSD at all. See the NBOMe warning a few pages on, because this is exactly where that danger lives.

**2. Did they know how much?** Eyeballed doses are the norm. A gram of mushrooms from one flush can be several times stronger than a gram from another. Without a scale and a tested product, the dose in the title is a guess.

**3. Was anything else involved?** Alcohol, cannabis and prescription medication appear in a great many reports but are mentioned in far fewer.

**4. What were the set and setting?** These drive outcomes enormously and are reported rarely. Somebody's state of mind going in, who they were with, and where they were will tell you more about what happened than the compound will.

**5. When was it written?** Most reports are written days or weeks later, and by then the story has been told a few times. Memory reshapes itself with each telling. Hofmann's journal is precious precisely because he was writing as it happened.

**6. Who is not writing?** This is the big one. The extraordinary experiences get posted and the ordinary ones do not. The person who had a pleasant, unremarkable afternoon rarely sits down to compose two thousand words about it. And the very worst outcomes are not represented at all, for obvious reasons.

**7. Is the language description or metaphor?** These experiences resist ordinary words, so people reach for images. "The walls were breathing." "I became the music." Those are honest attempts at the impossible, not literal reports. Read them as poetry doing its best, because that is what they are.

## THE MOST IMPORTANT PART

For decades, people writing about their trip experiences kept saying the same things. That they had stopped wanting to drink. That cigarettes had simply lost their hold. That a fear they had carried their whole lives had loosened. Formal science could not investigate any of it, so it sat in the archive, waiting patiently for the light to shine.

Then the labs reopened, and started checking.

**Smoking.** A Johns Hopkins pilot study of psilocybin combined with cognitive behavioral therapy found that sixty-seven percent of participants were still not smoking twelve months after their quit date. A subsequent randomized trial found fifty-nine percent verified abstinence at twelve months against twenty-three percent for nicotine replacement. Standard treatments do not come close to those numbers.

**Alcohol.** A randomized controlled trial published in *JAMA Psychiatry* in 2022 found that two doses of psilocybin alongside psychotherapy significantly reduced heavy drinking days in people with alcohol use disorder, compared against an active placebo.

**Opioids.** The ibogaine work described in Chapter Three began with one nineteen-year-old in 1962 who noticed his craving was gone and spent thirty years insisting somebody look into it.

**And the oldest fear, retired.** The 1960s claim that these substances cause lasting mental illness has not survived scrutiny. A population study of 130,152 American adults published in 2013 found no association between lifetime psychedelic use and mental health problems. That does not mean there is no risk. It means the alarm that closed the door in 1970 was not supported by the evidence.

Every one of those findings started as somebody typing up what happened to them and posting it where a stranger might read it.

The reports were never proof. But over and over, they were the ones pointing at the door, patiently, for years, until science finally came along with a key.

Read them carefully and with gratitude, knowing that somewhere in that archive right now is the next thing we are about to find out.

Facts first. And curiosity ALWAYS.

## THE STRAIGHT DOPE

Here is the single most important safety fact in this entire book.

LSD is active in micrograms, which means an effective dose fits on a scrap of paper. That has an ugly consequence: any potent compound can be sold on a piece of blotter paper and look exactly like LSD.

A family of compounds called **NBOMes**, particularly 25I-NBOMe, emerged around 2010 and were widely sold as LSD. They are not LSD. They are considerably more toxic, carry a much narrower margin between active and dangerous, and have killed a significant number of young people, many of whom believed they had taken something with an extremely good safety record. Very sad.

Actual LSD has no known lethal dose in humans. The blotter tab of paper you cannot verify absolutely does.

Reagent test kits are legal, inexpensive and available online. These kits can easily and quickly rule out dangerous fakes. This is not a small point. It is the difference between a substance with a remarkable safety profile and a genuine death lottery.

Facts first.

---

## WHAT COMES NEXT

Two frontiers, and one honest question.

**Ibogaine clinics.** We covered iboga in Chapter Three. The synthetic and semi-synthetic derivative ibogaine is now the basis of a treatment industry operating in a regulatory gray zone. Results for opioid dependence continue to impress, and the cardiac risk remains real. The Texas state funding for ibogaine research may finally bring this into the light.

**The trip-free psychedelic.** This is the most interesting question in labs right now.

Researchers at UC Davis and elsewhere have engineered compounds that appear to produce the neuroplastic effects of psychedelics, the synapse growth and rewiring that may drive the lasting antidepressant benefit, without producing hallucinations. One is **tabernanthalog**, derived structurally from ibogaine but engineered to be non-hallucinogenic and non-cardiotoxic.

If it works, the implications are enormous. A pill with all the benefits and none of the eight-hour sessions, therapist costs, and difficult experiences.

**And here is the honest question.** Is the *experience* the medicine, or is it a side effect?

Some researchers argue the mystical experience itself is doing the therapeutic work, and point to consistent findings that the intensity of that experience predicts how well patients do afterward. Strip it out and you may have stripped out the whole point.

Others argue that is romanticism, that the biology is the biology, and that requiring patients to endure eight difficult hours when a pill would do is neither compassionate nor scalable.

Both sides have real arguments. Neither has won. But I suspect we will eventually discover that one is right and one is wrong.

We started this chapter ten thousand years from a mushroom in a pasture, and we end it in a laboratory trying to decide whether to remove the very thing every one of those ten thousand years was after.

Where do you land on that? It is genuinely one of the most interesting and consequential questions anyone is asking right now.

CHAPTER EIGHT

## The Synthetic Century

*From a forgotten Merck patent to an FDA priority review voucher*

**THE FIRST HALF**

- 1912 ◆ MDMA patented by Merck
- 1938 ◆ LSD-25 synthesized
- 1943 ◆ Bicycle Day
- 1947 ◆ Sandoz markets Delysid

**THE MIDDLE YEARS**

- 1953 ◆ MK-Ultra begins
- 1962 ◆ Ketamine synthesized
- 1970 ◆ Controlled Substances Act
- 1976 ◆ Shulgin rediscovers MDMA

**THE RETURN**

- 1985 ◆ MDMA scheduled
- 1986 ◆ MAPS founded
- 2000 ◆ Ketamine's antidepressant effect
- 2019 ◆ Esketamine approved
- 2024 ◆ MDMA complete response letter
- 2026 ◆ National priority vouchers

How the World Gets High · TJ Toce

CHAPTER EIGHT

## The Molecular Family Tree

*How everything in this book is related · and three things that are not**

**THE CLASSICAL PSYCHEDELICS**

*all three families act on the same lock: the 5-HT2A serotonin receptor*

| TRYPTAMINES<br>*built on the indole ring, like serotonin* | PHENETHYLAMINES<br>*built on a benzene ring, like dopamine* | LYSERGAMIDES<br>*built on ergoline, from a rye fungus* |
|---|---|---|
| Psilocybin / Psilocin<br>*fungi* | Mescaline<br>*peyote, San Pedro* | LSD<br>*Hofmann, 1938* |
| DMT<br>*plants, and you* | MDMA<br>*synthetic* | LSA / ergine<br>*morning glory* |
| 5-MeO-DMT<br>*the toad* | 2C-B, 2C-I, 2C-E<br>*Shulgin's family* | Ergotamine<br>*migraine medicine* |
| Ibogaine<br>*iboga root* | Methylone<br>*in trials, 2026* | Ergometrine<br>*obstetrics* |
| Bufotenine<br>*toad, seeds* | Myristicin<br>*nutmeg* | |

*** THREE THAT BELONG TO NO FAMILY HERE**

**Ketamine** *an arylcyclohexylamine, acts on NMDA glutamate receptors*
**Salvinorin A** *a diterpene, acts on the kappa-opioid receptor*
**Muscimol, from the fly agaric** *an isoxazole, acts on GABA-A receptors*

*Not one of them is a classical psychedelic. They are in the book. They are simply not on this tree.*

How the World Gets High · TJ Toce

# CHAPTER NINE

## 2026 AND BEYOND – The Renaissance Has Begun

*How psychedelics are making their way back into the mainstream of society once again*

---

> *One resists the invasion of armies; one does not resist the invasion of ideas.*
>
> **VICTOR HUGO** *Histoire d'un crime, 1877*

---

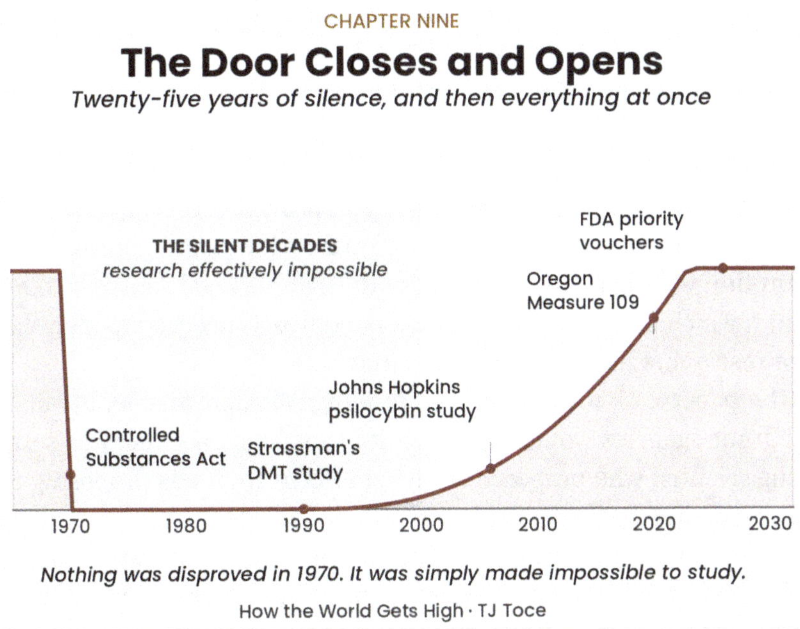


---

Here is the thing that nobody in this field predicted, and I mean nobody.

Ask a psychedelic researcher in 2010 when psilocybin might become an approved medicine and you would have gotten a careful, hedged, deeply cautious answer involving the word "decades" and a lot of throat-clearing about political climate.

Ask that same researcher today and the honest answer is that it could be a matter of months.

The door that slammed shut in October 1970 has not merely reopened. It has come off its hinges. And it happened faster than the people who spent their careers pushing on it ever allowed themselves to hope.

Let's walk through exactly how we got here and other potentialities.

---

### THE SILENT DECADES

From 1970 to roughly 1990, serious clinical psychedelic research in the United States essentially did not exist.

It is worth being precise about why, because the reason is not simply that the drugs were banned. Plenty of Schedule I and II substances are studied. The reason is that Schedule I created a bureaucratic wall that made research practically, rather than legally, impossible. That's another way of saying RED TAPE!

To study a Schedule I compound a researcher needed a special DEA registration, a specific and separately approved protocol, a secure safe with documented access logs, institutional review board approval from an institution willing to be associated with it, and a legal source of the compound, which barely existed because nobody was licensed to manufacture it for research. Can you say vicious circle?

The paperwork took years. The funding was unavailable, because federal grant agencies would not touch it. And the career risk was severe. A young scientist who proposed studying LSD in 1978 was proposing to end their own career.

So an entire branch of psychiatry sat frozen in a drawer while medicine transformed around it. We sequenced the human genome. We built machines that watch a brain think in real time. And the most promising results for treating alcoholism sat untouched since 1960.

Twenty-five long years. Consider what that cost.

---

## PLAYING THE LONG GAME

The reopening did not happen by accident. It happened because a small number of visionary people spent their entire adult lives working on it. The easiest way for one person to move a huge boulder is to break it into small pebbles and move them one by one.

The Way-Showers:

**Rick Doblin** founded MAPS in 1986 at twenty-nine, with the stated goal of making MDMA a legal prescription medicine. He then earned a doctorate in public policy from Harvard specifically to understand how to navigate the FDA. He has been at it for forty years.

**Rick Strassman** spent two years getting approval for the DMT study we covered in Chapter Seven, and when it began in 1990 it was the first new human psychedelic research in the country in over two decades. He proved it could be done.

**Roland Griffiths** at Johns Hopkins was a rigorously mainstream researcher with an impeccable reputation in drug abuse science. That reputation was the point. When he finally turned his attention to psilocybin, he brought credibility nobody could dismiss.

**2006** is the year everything changed. Griffiths and colleagues published a study in which healthy volunteers received psilocybin under supportive conditions. Two thirds rated it among the five most personally meaningful experiences of their lives, comparable to the birth of a child. At a fourteen-month follow-up, they still did. That's huge!

It was published in a respected journal. The methodology was airtight. And it was, quietly and effectively, an announcement that this field was scientifically respectable again.

The Johns Hopkins Center for Psychedelic and Consciousness Research was formally launched in 2019 with private funding from this amazing group of people we all owe a huge debt of gratitude to:

**Steven & Alexandra Cohen Foundation; Tim Ferriss** (author and technology investor); **Matt Mullenweg** (co-founder of WordPress); **Blake Mycoskie** (founder of TOMS); **Craig Nerenberg** (investor)

Imperial College London opened its own center the same year. Then NYU, UCSF, Mount Sinai, Yale and dozens of others followed down the golden path.

The dam did not break. It was disassembled, stone by stone, by people who were extremely patient and saw the future.

## THE STATES POINTED THE WAY

While federal researchers worked through the FDA, something else was happening at the state level, and it moved faster than expected.

**OREGON.** In November 2020 voters passed Measure 109, making Oregon the first jurisdiction in the modern world to create a legal, regulated framework for supervised psilocybin use.

It is not a dispensary model where you can buy psilocybin and take it home. It is not a medical model either, because no diagnosis is required and no doctor is involved.

What it is: you meet with a licensed facilitator for a preparation session. You consume psilocybin at a licensed service center. The facilitator stays with you for the entire experience, typically around six hours. You have an integration session afterward.

The first center opened in June 2023 in Eugene, sitting among the vineyards of the Willamette Valley. Thousands of sessions have now taken place. Much-needed modifications to the program were signed into law in March 2026 to make it more sustainable.

**COLORADO.** Voters passed Proposition 122 in November 2022 and the legislature built the framework with Senate Bill 23-290 in 2023.

Colorado went considerably further than Oregon in two ways. Its licensed healing center model is similar, but it ALSO permits personal use, personal cultivation and sharing without payment for adults twenty-one and over. And it covers not just psilocybin and psilocin but DMT, ibogaine and mescaline, making it the broadest framework in the country and sitting at the vanguard.

Thankfully, Colorado deliberately excluded peyote, at the explicit request of Native American leaders, for the conservation reasons we discussed in Chapter Three. That was totally the right call, and it deserves a shout out!

**NEW MEXICO.** Senate Bill 219, the Medical Psilocybin Act, was signed in April 2025, making New Mexico the third state and the first to go through a legislature rather than a ballot initiative.

New Mexico chose a medical model. Qualifying conditions, physician involvement, and a treatment fund to help patients afford it. First patient enrollment is targeted for late 2026.

Three states. Three genuinely different models. And we are all about to learn a great deal from watching them run side by side in the national spotlight.

---

## 2026: THE YEAR IT ACCELERATED

At last we arrive at the here and now, and the pace of the last several months has been exhilarating.

**April 18, 2026.** President Trump signed **Executive Order 14401**, "Accelerating Medical Treatments for Serious Mental Illness," directing the FDA to prioritize review of psychedelic compounds carrying Breakthrough Therapy designation.

Whatever your politics, this change is a big deal! Psychedelic medicine has become a genuinely bipartisan issue, driven substantially by veterans' advocacy, and that cross-party support is the single biggest structural change in this field's political position in fifty years.

**April 24, 2026.** The FDA issued **National Priority Vouchers** to three developers:

1. **Compass Pathways**, for COMP360, a synthetic psilocybin formulation, targeting treatment-resistant depression.
2. **The Usona Institute**, a nonprofit, for psilocybin targeting major depressive disorder.
3. **Transcend Therapeutics**, for methylone, an MDMA-adjacent compound, targeting PTSD.

The voucher significantly compresses the FDA review timeline to one or two months following submission, against a standard review of 10-12 months minimum.

Compass has indicated a rolling New Drug Application submission targeting the fourth quarter of 2026, which places a potential approval decision in late 2026 or early 2027.

**July 2026.** The FDA published its **final guidance on psychedelic clinical investigations**, which sounds like paperwork but is actually an enormous move.

Here is why. Every psychedelic trial has faced the same methodological problem, called functional unblinding. In a normal drug trial, neither the patient nor the researcher knows who got the drug. With psychedelics, everybody knows within an hour. That was the central criticism leveled at the MDMA application in 2024.

The FDA's final guidance provides a desperately needed official framework for how to handle it. Which means the field now has a rulebook, written by the regulator, telling developers exactly what an acceptable trial looks like.

For a field that spent fifty years being told no, being handed a rulebook is the single most encouraging development of all.

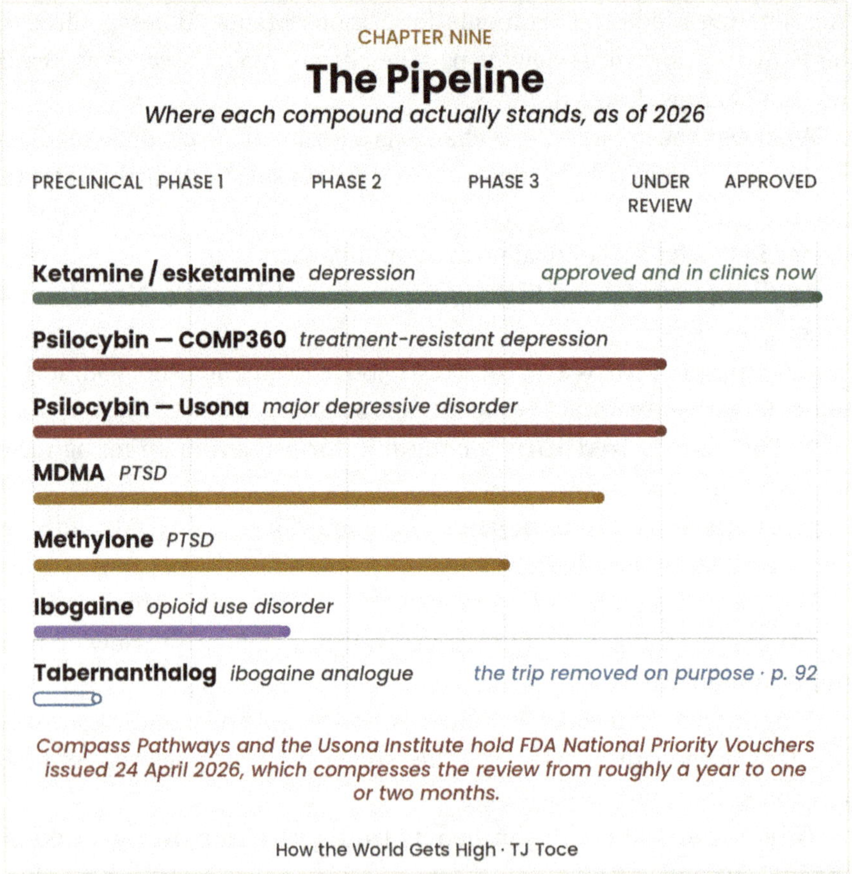

## WHY VETERANS CHANGED EVERYTHING

If you want to understand why this became politically possible, thank your veterans.

Roughly 7% of American veterans will experience PTSD in their lifetime, and among those who served in Iraq and Afghanistan the lifetime rate approaches 29%. Existing treatments help many people and fail a great many others. Tragically, the suicide rate among veterans has remained a national emergency for two decades.

Veterans began seeking out ibogaine and 5-MeO-DMT at clinics in Mexico, largely on their own initiative and their own money, and they came back and told people. Loudly.

Stanford researchers published a study of veterans with traumatic brain injury treated at a Mexican ibogaine clinic and reported substantial improvements in PTSD, depression, anxiety and cognitive function.

Then veterans did what veterans do. They organized. They testified before state legislatures. They walked into congressional offices, and they were extremely hard to dismiss, because a decorated combat veteran saying "this saved my life and I had to leave the country to get it" is an argument that cuts across every ideological line in American politics.

Texas approving ibogaine research funding in 2025 happened because of veterans. Executive Order 14401 happened, substantially, because of veterans.

The Department of Veterans Affairs is now funding psychedelic research directly, the first federal funding of its kind in more than half a century. Thank your veterans!

---

## THE REST OF THE WORLD, AND SOME VERY STRANGE STORIES

We are not alone in this. In several respects we are behind. And the international picture contains some of the most delightful, absurd and genuinely moving stories on this subject.

Let's take a ride.

## THE COUNTRY THAT STUDIED IT AND SAID YES

Start with the best one, because almost nobody in North America has heard it. Deliberately!

In the mid-1980s the Brazilian government had a problem. Two ayahuasca churches, Santo Daime and the União do Vegetal, had grown from small Amazonian congregations into national organizations with thousands of members, openly drinking a beverage containing active DMT.

The obvious move was prosecution. Brazil did something else entirely.

The federal drug council, CONFEN, convened a working group. And rather than reviewing paperwork from an office in Brasília, that working group went out and LOOKED. They traveled the country, visited the churches and observed the brew being made. They spent time in the communities and talked to the people living in them.

The investigation ran for two years.

The report landed in September 1987, and its conclusion is one of the most remarkable sentences any government has ever written about a psychedelic. The working group found that the religious rituals conducted with the sacramental drink brought no injury to social life, and further, that they appeared to CONTRIBUTE to better social integration, with the benefits attested by the members themselves.

Read that again. A national drug commission spent two years investigating a psychedelic church and concluded that it was making its members better citizens. And why haven't you heard of this before?

Logically, the plants were kept off the prohibited list.

This stands as the clearest example anywhere on Earth of a government asking a question, doing the work to answer it, and then following the evidence wherever it led. Unbiased!

Everybody else banned first and asked later. Brazil went and had a look.

### THE COUNTRY THAT PROTECTS ITS PLANT

**GABON** does something no other nation does.

Iboga, the root from Chapter Three, is not merely tolerated there. It is treated as a national patrimony. Bwiti is a recognized religion practiced by a substantial share of the population, and iboga sits at the center of it.

And because global demand for ibogaine has grown fast enough to threaten wild populations, Gabon has restricted export and moved to control harvest.

Think about the reversal. Most of the world's conversation about psychedelics is about protecting people from the plant. Gabon's is about protecting the plant from the world.

Given what happened to peyote in Texas, and what is happening to the Sonoran Desert toad right now, Gabon may be the only country that has its priorities in the right order.

### THE EXEMPTION HIDING IN MEXICAN LAW

Here is one that ties directly back to Chapter One, and it deserves a closer look.

Psilocybin is prohibited in Mexico under federal law. But the Federal Penal Code contains a provision exempting quantities used in traditional ceremony and custom by indigenous peoples and communities where such use is established practice.

Which means that the velada, the Mazatec night vigils and the ceremony María Sabina conducted her entire life, is legally protected in Mexico today.

It is imperfectly enforced and imperfectly understood, and it does not stop the tourist industry that grew up in Huautla de Jiménez. But the legal principle is there, in statute, and it says that a tradition older than the Spanish conquest has a right to continue.

After what this book told you happened to María Sabina, it is worth knowing that her people's practice is protected in her own country.

### JAMAICA, WHICH SIMPLY NEVER GOT AROUND TO IT

Jamaica's status is the strangest of all, because it is not a reform. It is an omission.

When Jamaica implemented its controlled substances framework, psilocybin was never scheduled. Not decriminalized, not legalized, not debated and permitted. Simply never added to the list.

The result is that Jamaica is one of the very few places on Earth where psilocybin mushrooms are, in a plain sense, legal, and a substantial retreat industry has grown up on that basis, particularly around Negril and the north coast.

Somewhere in the history of Jamaican legislative drafting is a list that somebody did not finish. Fifty years later it supports an industry probably enlightening a lot of people.

### PERU MADE IT NATIONAL HERITAGE

In 2008, Peru's National Institute of Culture declared the traditional use of ayahuasca to be part of the nation's **cultural heritage**.

Not merely legal. Not merely tolerated. Officially recognized as something Peru considers essential to its own identity, on the same footing as its archaeological sites and its traditional music. Talk about way-showers!

The declaration also contained a warning that reads like a direct response to Chapter One. It expressed concern about decontextualized commercial use, and about the brew being separated from the traditional framework that gives it meaning. Sacred matters.

Peru saw what happened at Huautla and tried to get ahead of it. Beautifully.

### THE ONES DOING IT PROPERLY

**AUSTRALIA**, **CANADA**, **SWITZERLAND**, **THE CZECH REPUBLIC** and **THE NETHERLANDS** were all covered earlier in the book.

### AND NOW FOR THE ONES GOING THE OTHER WAY

Let's highlight a few counterexamples sadly making a scene.

**FRANCE** moved decisively in the opposite direction of the progressive countries above. In 2005, following a court case, France classified not just DMT and harmala alkaloids but the PLANTS themselves, listing *Banisteriopsis caapi*, *Psychotria viridis* and related species as controlled. It remains one of the most restrictive positions in Europe.

**ITALY** banned kambo for practitioner use following a fatality.

This movement does not win every vote and still has a road ahead.

### WHAT THE WORLD MAP ACTUALLY SHOWS

Step back and look at the whole thing and a pattern emerges worth shouting out.

The countries with the most settled, least anxious relationships to these substances are, almost without exception, the ones where the practice never fully stopped. Brazil. Peru. Gabon. Mexico. Places where a living tradition survived the twentieth century and gave the modern conversation something to stand on.

The countries in the most confusion are the ones where the thread was cut. Where a substance arrived with no cultural container, no elders, no protocol, no accumulated wisdom about how to hold it, and so the only available response was prohibition.

Which suggests something rather hopeful about where we are heading.

We are not, in the end, trying to invent something new. We are trying to rebuild a "container" that most of the world never lost.

And there is one place under the American flag where that "container" was carried across four thousand miles of open ocean, in a canoe, and never set down. One of my favorite places on the planet.

Let's go there next.

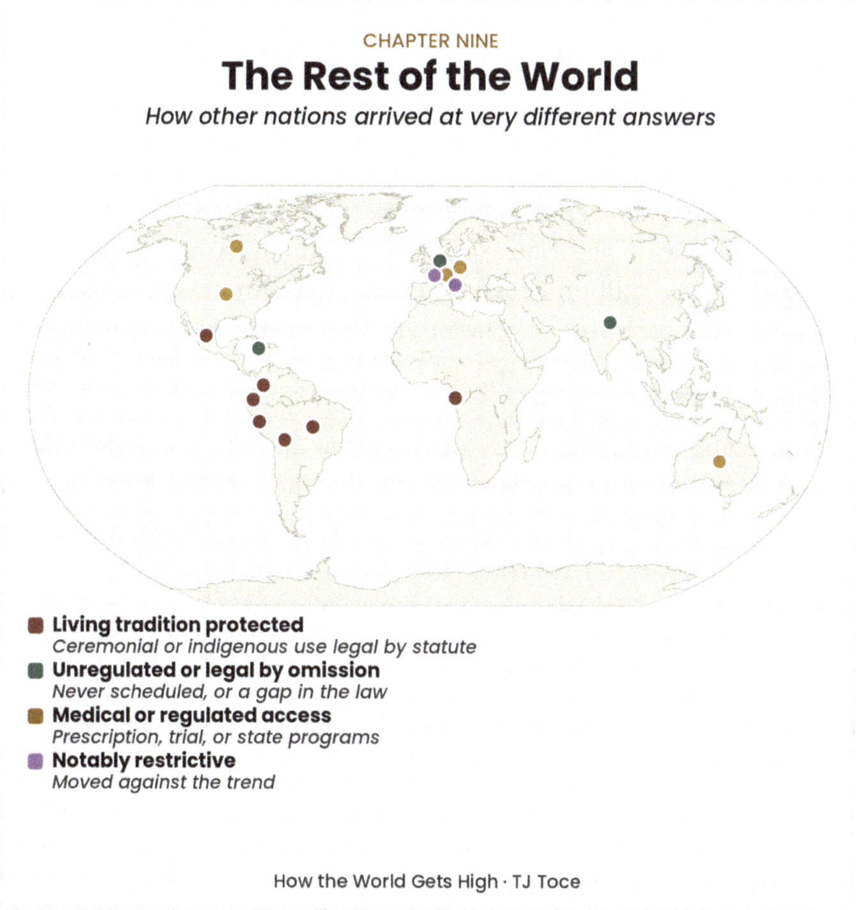

## NOW THE HARD PART

This book promised you honesty over enthusiasm, and a chapter this optimistic owes you a peek at the other side of the ledger.

**1. The hype is running well ahead of the evidence.** Psychedelics are being described in popular media as a cure for depression, addiction, PTSD, anxiety, eating disorders, obsessive-compulsive disorder and existential distress. The actual evidence is genuinely promising but considerably narrower.

**2. Commercialization is a genuine risk.** Substantial venture capital has flowed in. Companies have filed patents on things that arguably should not be patentable, including protocols, room designs and naturally occurring compounds in slightly modified form. The tension between the nonprofit and open-science wing of this field is palpable.

**3. Access is still not equal.** Without deliberate policy intervention, this is currently a treatment for people who already have resources, involving conditions that fall hardest on people who do not.

**4. Reciprocity remains unresolved.** Nearly everything in this book came from indigenous knowledge. The Mazatec. The Amazonian peoples. The Bwiti. The Native American Church. A multibillion-dollar industry is being built on knowledge those communities developed, preserved through centuries of persecution, and shared without being compensated.

María Sabina died very poor.

Organizations like the Chacruna Institute and the Indigenous Medicine Conservation Fund are working on this. But the honest status is that the question remains open, and the industry is moving considerably faster than its conscience. You can change that.

---

## WHERE THIS GOES

So what actually happens next? Here is the reasonable expectation.

**In the next year or two**, an FDA approval for psilocybin in treatment-resistant depression is genuinely likely. That would be the first approval of a classical psychedelic as a medicine in American history, and it would change the legal and cultural conversation permanently.

**In the next five years**, expect more states to build access frameworks, insurance coverage fights, a serious shortage of trained facilitators, at least one significant scandal, and a hard public argument about whether the trip-free compounds from Chapter Eight represent progress or a retreat from what Mother Nature intended.

**In the next ten years**, the interesting question is not medical at all. It is what happens when a society that spent fifty years calling these substances dangerous and worthless has to explain to itself why it was so wrong for so long. But the positives here greatly outweigh the negatives. So, move on.

This is going to be fascinating to watch.

---

## THE LONGER VIEW

Let's recap the timeline.

Nine thousand years ago, somebody painted a running figure with mushrooms sprouting from his arms onto a rock face in a green Sahara.

Five thousand seven hundred years ago, somebody in what is now south Texas dried peyote buttons and left them in a cave.

Three thousand two hundred years ago, somebody carved a figure holding a San Pedro cactus into a stone wall at Chavín de Huántar.

Two thousand four hundred years ago, ten thousand Greek soldiers ate the wrong honey in a Turkish mountain village.

One thousand years ago, somebody in the Bolivian Andes filled a bundle made of fox snouts with the ingredients of ayahuasca.

Four hundred and seventy years ago, a Franciscan friar wrote down what he saw and disapproved of in Mexico.

Seventy-one years ago, a banker sat in a dark room in Oaxaca and heard a Mazatec healer sing.

Eighty-three years ago, a Swiss chemist rode a bicycle home.

Fifty-six years ago, a door slammed shut.

And this year, in 2026, a regulator in Maryland is reading an application.

Fifty-six years of prohibition, set against ten thousand years of continuous human practice, looks less like the natural order of things and much more like what it actually was. A brief and strange interruption.

The interruption is finally ending. The long story is picking back up where it left off, and this time we have brain scanners and clinical trials and regulatory frameworks and randomized controls to go alongside the drums and the singing.

That is not a replacement for what came before. It is the same question, asked with better instruments and a new point of view.

The running man on the rock face has been waiting nine thousand years for someone to catch up.

We are finally getting close. We are indeed ready.

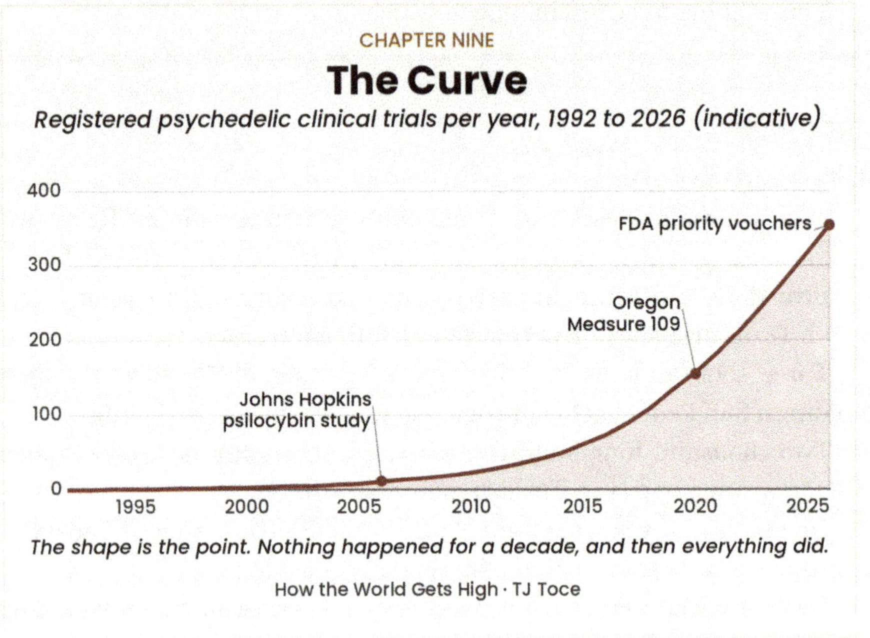


*The shape is the point. Nothing happened for a decade, and then everything did.*

# CHAPTER TEN

## A HAWAIIAN INTERLUDE

---

*No alien land in all the world has any deep, strong charm for me but that one, no other land could so longingly and so beseechingly haunt me, sleeping and waking, through half a lifetime, as that one has done. Other things leave me, but it abides; other things change, but it remains the same.*

**MARK TWAIN  Speech at Delmonico's, New York, 8 April 1889**

---

Roughly sixteen hundred years ago, give or take, a group of people got into voyaging canoes, great outriggers, somewhere in the Marquesas or the Society Islands and pointed north into open ocean.

They had no instruments. They navigated by stars, by swell patterns, by the flight paths of birds and the color of clouds over unseen land. And they were not fleeing anything. They were LOOKING, deliberately, for islands they had reasoned must be out there.

They found Hawai'i.

Now here is the part that matters for this book. Consider what you would pack.

You have limited space in a canoe crossing thousands of miles of empty Pacific. Every cutting, root and seedling you bring costs you weight, water and room for people. You have to choose what a civilization cannot be rebuilt without.

The Polynesian voyagers brought about twenty-four species. They are called the **canoe plants**, and they are the founding botany of Hawaiian civilization. Kalo, the taro that became poi. 'Ulu, breadfruit. Mai'a, banana. Kō, sugarcane. Niu, the coconut. Wauke, the paper mulberry for making kapa cloth. Ti. Sweet potato. Turmeric.

And among the very few non-food plants they judged essential enough to carry across an ocean:

**'Awa.**

## THE PLANT THAT MADE THE CUT

'Awa is *Piper methysticum*, a member of the pepper family, known across the wider Pacific as **kava**, or kava kava.

The root is pounded or chewed, mixed with water, strained, and drunk from a shared bowl. The effect arrives within minutes: the mouth and tongue go numb, the muscles let go, anxiety recedes, and what remains is a calm, sociable, clear-headed ease. People describe it as the tension leaving the body while the mind stays entirely present.

**And now a point of clarity since this is a book about psychedelics.**

Kava is not a hallucinogen. It is frequently described as one, including in a great deal of otherwise reliable writing, and the description is simply wrong.

The active compounds are **kavalactones**, principally kavain, dihydrokavain, methysticin, dihydromethysticin, yangonin and desmethoxyyangonin. They work through a genuinely fascinating multi-target mechanism, modulating GABA-A receptors while also touching dopaminergic, serotonergic and glutamatergic pathways and inhibiting MAO-B.

That is a sophisticated piece of pharmacology and it produces a real, reliable, valuable effect. It is an anxiolytic and a muscle relaxant of the first order. The original cocktail!

What it does not do is produce visions, alter perception, dissolve the self or open any door of the kind the rest of this book describes. You are entirely yourself the whole time, feeling considerably better than you did an hour ago.

So why does 'awa belong in this book at all?

Because the question this book is really asking is not "what makes you hallucinate." It is what humans reach for when they want to change how they feel, and what they describe as being high!

By that measure 'awa is one of the great answers in human history. And what the Hawaiians built around it is extraordinary. Facts first, and then the wonder.

## THIRTEEN CULTIVARS AND A DEATH PENALTY

Here are some curious details.

'Awa does not reproduce from seed. Every plant is a cutting of another plant, which means every 'awa plant alive in Hawai'i today is, in a real physical sense, a continuation of something a voyager carried in a canoe sixteen centuries ago.

We came across this before in this book, in the Sierra Mazateca, with *Salvia divinorum*. Two sacred plants, on opposite sides of the Pacific, both kept alive for centuries entirely by human hands, one cutting at a time, because they cannot do it themselves.

Old Hawai'i recognized more than a dozen named varieties. Thirteen cultivars survive today, and they were not interchangeable.

**Mō'ī**, "the king," is unusually rich in kavain. It was reserved for royalty and priests, and commoners were forbidden to drink it. The prohibition was not a suggestion. Accounts hold that the penalty for violating it was death.

**Hiwa** produces a brew so dark it runs nearly black. This was the variety of the kahuna, the priests, and it was the 'awa of divination. When a kahuna needed to see something, this is what he drank.

**Kumakua**, **Nene**, **Papa** and the others each carried their own use, their own occasion, their own standing.

An entire pharmacopeia, encoded in a plant that cannot reproduce without you, carried across an ocean and then differentiated over centuries into thirteen distinct strains with distinct social meanings.

That is not a beverage. That is a civilization's sacred marriage with a plant.

---

## WHAT IT WAS FOR

'Awa sat at the center of Hawaiian life in a way that has no clean Western equivalent.

It was drunk at the coronation of chiefs. It was drunk at the naming ceremonies of infants. It was offered to the gods, poured out as a sacrament, and offered to honored visitors as the highest expression of hospitality. Kahuna used it in divination and in healing. Fishermen and farmers drank it after work in a way that would be recognizable to anyone who has ever met a friend for a beer, except that the plant they were drinking had been carried there on purpose by their ancestors.

Then came the nineteenth century, and missionaries, and the systematic suppression of Hawaiian practice, and ʻawa went the way of hula and the Hawaiian language. Discouraged. Diminished. Not extinguished.

And like hula and the language, it has come back.

ʻAwa cultivation is expanding across the islands. Kava bars operate in Honolulu, Hilo and Kailua, serving it in traditional coconut shell bowls to people who mostly have no idea they are participating in something sixteen hundred years old. In 2025, Hawaiʻi formally affirmed kava's status as a safe traditional food.

The plant that made the cut for the canoe is flourishing. There is something deeply satisfying about that. It's a good thing.

A HAWAIIAN INTERLUDE

## The Canoe Plants

*What a civilization packs when it cannot come back for more*

| | |
|---|---|
| Kalo (taro) | 'Ulu (breadfruit) |
| Mai'a (banana) | Kō (sugarcane) |
| Niu (coconut) | 'Uala (sweet potato) |
| Wauke (paper mulberry) | Kī (ti) |
| 'Ōlena (turmeric) | Uhi (yam) |
| Pia (arrowroot) | Ipu (bottle gourd) |
| Hau | Kukui (candlenut) |
| Milo | Kamani |
| Noni | Ōhi'a 'ai (mountain apple) |
| Ko'oko'olau | Ipu 'awa'awa |
| Ape | Hoi |
| Olonā | 'AWA (kava) |

*Twenty-four species judged worth the space on a canoe crossing four thousand miles of open ocean.*

*'Awa was one of the very few that was not food.*

How the World Gets High · TJ Toce

---

### WHAT ARRIVED LATER, AND NOT BY INVITATION

'Awa came deliberately, chosen, packed and hand-carried. Everything else in this chapter arrived by accident.

**Hawaiian baby woodrose.** *Argyreia nervosa* carries the name but is not Hawaiian at all. It is native to the Indian subcontinent, arrived as an ornamental, escaped, and naturalized so thoroughly across the islands that the world now calls it Hawaiian. Its seeds contain LSA, the compound from Chapter Three that Albert Hofmann found in Mexican ololiuqui. A vine from India, named for Hawai'i, carrying a molecule the Aztecs used.

**Mushrooms.** Hawai'i's climate is close to ideal. *Psilocybe cubensis* thrives in cattle pasture on the Big Island. *Panaeolus cyanescens*, a small, potent, blue-bruising species, is so abundant in Hawaiian and other tropical pastures that "Hawaiian" is a common name attached to it in cultivation. Neither is native. Both followed cattle, exactly as Chapter Two described, riding along with agriculture the way they have everywhere else on Earth.

**The ayahuasca vine.** *Banisteriopsis caapi* grows well in Hawai'i and is cultivated there, which means an Amazonian liana now climbs on Pacific islands ten thousand miles away.

**Datura and brugmansia** are naturalized across the islands, growing in gardens and along roadsides, as beautiful and as unwelcome as Chapter Three described them.

These islands became one of the more remarkable botanical crossroads on the planet.

---

## AND WHERE HAWAI'I IS HEADED

Hawai'i has not legalized or decriminalized anything yet but is moving, and in a manner that is very much its own. Kama'āina style.

A state psychedelics task force first convened in 2023. And in the 2026 session, the Hawai'i Senate passed SB 3199, establishing a new task force charged with studying and recommending pathways to therapeutic access for psilocybin and MDMA, involving several governmental departments and the University of Hawai'i medical school. It raced through House committees in the spring.

The vote in the Senate was 24 to nothing! Aloha!

No state gets to a unanimous vote on this subject by accident. Hawai'i is a sacred place with a living, continuous, publicly celebrated tradition of a plant used ceremonially in a shared bowl, and where the idea that a plant might sit at the center of healing and community is simply taken for granted.

The "container" was never lost here. It came ashore in a canoe sixteen hundred years ago and it has been passed around ever since.

Everywhere else in this book, we have watched people trying to rebuild something.

In Hawai'i, they just kept it.

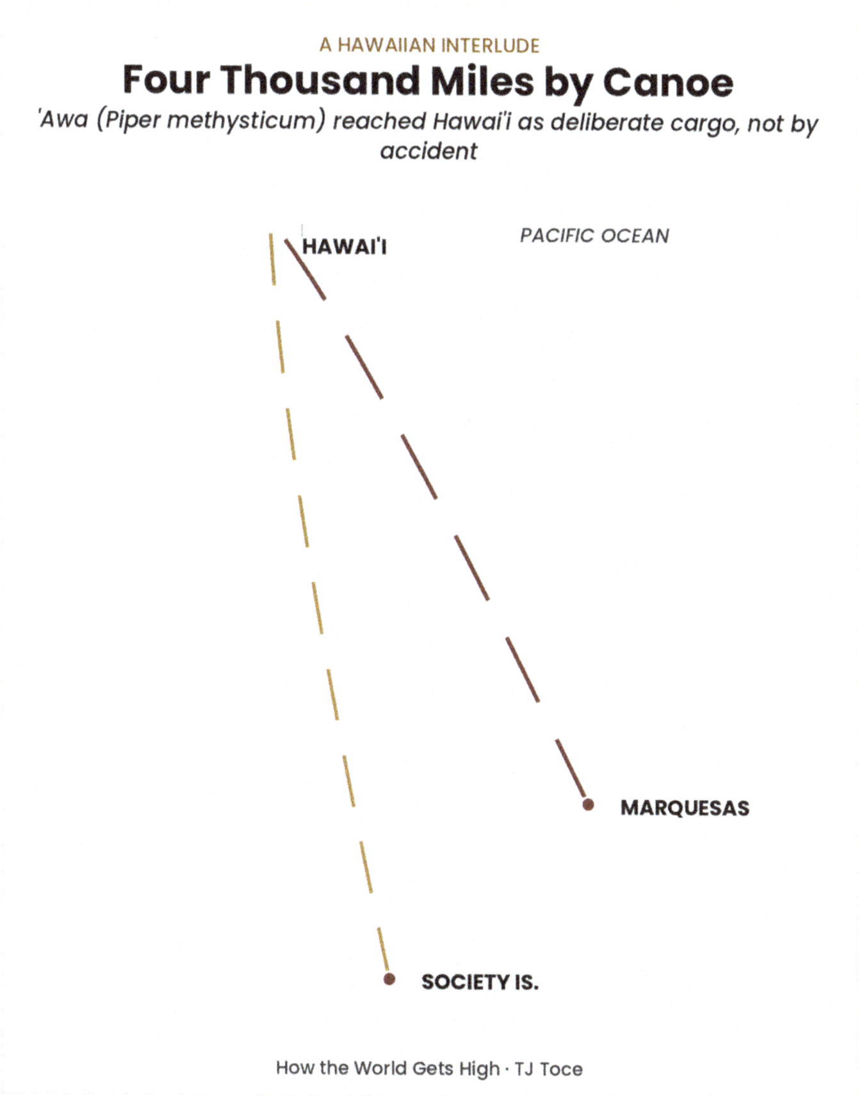

# CHAPTER ELEVEN

## THE MAP REDRAWN

*Cities and towns in the United States that have legalized or decriminalized*

> *Laws and institutions must go hand in hand with the progress of the human mind. As that becomes more developed, more enlightened, as new discoveries are made, new truths discovered and manners and opinions change, with the change of circumstances, institutions must advance also to keep pace with the times.*
>
> **THOMAS JEFFERSON Letter to Samuel Kercheval, July 12, 1816**

CHAPTER ELEVEN

# The Map Redrawn

*United States psychedelic policy as of 2026. In May 2019 this map was entirely blank.*

- **Regulated access program (OR, CO)**
- **Medical program (NM)**
- **City or county measure (~30)**

*Decriminalization and deprioritization do not make possession legal. Federal law is unchanged everywhere on this map.*

How the World Gets High · TJ Toce

---

Before you read a single entry in this chapter, read this section. I know we covered it earlier, but it is very important and getting it wrong is how people get arrested.

---

## THREE WORDS THAT ARE NOT THE SAME THING

People use these three terms interchangeably. They are not interchangeable, and the difference between them is the difference between a fun afternoon and a criminal record.

**1. LEGALIZATION.** A regulated system exists made of licenses, rules, locations and a lawful supply chain. This is the only category where something is affirmatively permitted.

**2. DECRIMINALIZATION.** Penalties are removed or reduced, usually converting a criminal offense into a civil one, similar to a parking ticket. Possession is still not permitted. There is no legal way to buy it.

**3. DEPRIORITIZATION.** This is what almost every city on the list below actually did. The local government passed a resolution or ordinance directing police to treat these offenses as their lowest enforcement priority. **The conduct remains fully illegal.** The city has simply asked its officers not to focus on it. I can see you rolling your eyes.

And now the fact that overrides all three:

**FEDERAL LAW REMAINS UNCHANGED EVERYWHERE ON THIS LIST.**

Psilocybin, psilocin, mescaline, DMT and ibogaine remain Schedule I controlled substances under federal law in all fifty states, in every city named below, without exception. Federal enforcement against individuals for personal possession is rare.

A city council resolution cannot amend the Controlled Substances Act. Nothing in this chapter is legal advice, and nothing in it makes anything federally legal.

Facts first. Now let's look at the map.

CHAPTER ELEVEN

## Three Words That Are Not the Same Thing

*Confusing these is how people get arrested*

**LEGALIZATION** — *Oregon · Colorado · New Mexico*

A regulated system exists. Licenses, rules, lawful supply.

**DECRIMINALIZATION** — *Some measures*

Penalties reduced or removed. Possession still not permitted.

**DEPRIORITIZATION** — *Almost every city on the list*

Fully illegal. Police directed to treat it as lowest priority.

**FEDERAL LAW IS UNCHANGED EVERYWHERE**

Psilocybin, mescaline, DMT and ibogaine remain Schedule I in all fifty states.

How the World Gets High · TJ Toce

---

## PART I: WASHINGTON, D.C.

### Initiative 81 — Passed November 2020

The Entheogenic Plant and Fungus Policy Act of 2020 passed with roughly seventy-six percent of the vote, one of the strongest margins any psychedelic measure has received anywhere.

It directs Metropolitan Police to treat non-commercial cultivation, distribution, possession and use of entheogenic plants and fungi as among the lowest law enforcement priorities.

It is deprioritization, not legalization. And because the District's laws are subject to congressional review, its position is uniquely precarious.

---

## PART II: THE CITIES AND COUNTIES

Approximately thirty municipalities and counties across a dozen states. Organized by state, in order of first action.

Unless otherwise noted, every entry below is **deprioritization**, meaning the conduct remains illegal and the local government has directed that enforcement be a low priority.

### COLORADO

**Denver** — May 2019. Initiative 301, the first psilocybin measure to pass anywhere in the United States, by a margin of under two thousand votes. Now superseded by the statewide program.

### CALIFORNIA

**Oakland** — June 2019. The second city in the nation, and the first to cover all entheogenic plants and fungi rather than psilocybin alone. Passed unanimously. **Santa Cruz** — January 2020. Unanimous. **Arcata** — October 2021. **San Francisco** — September 2022. Resolution covering all entheogens. **Berkeley** — 2023.

### MICHIGAN

Michigan holds the second-largest cluster in the country. **Ann Arbor** — September 2020. Unanimous. **Washtenaw County** — January 2022. County Prosecutor declined to charge entheogen offenses countywide. **Detroit** — November 2021. Passed by ballot measure, Proposal E. **Hazel Park** — 2021. **Ferndale** — 2022. **Pontiac** — 2022. **Ypsilanti** — 2022.

### MASSACHUSETTS

The largest cluster in the country, with eight municipalities. **Somerville** — January 2021, the first in the state. **Cambridge** — February 2021. **Northampton** — March 2021. **Easthampton** — October 2021. **Amherst** — 2021. **Salem** — 2021. **Provincetown** — 2022. **Medford** — 2023.

*Note: A statewide Massachusetts ballot measure to create a regulated access program was defeated in November 2024, which is a useful reminder that this movement does not win every time.*

### WASHINGTON

**Seattle** — October 2021. Unanimous. The largest city in the country to act at that point. **Port Townsend** — 2021. **Olympia** — 2021. **Jefferson County** — 2022.

### MINNESOTA

**Minneapolis** — 2023. By mayoral executive order rather than council vote, directing police to treat personal possession and use as the lowest enforcement priority.

### OTHER JURISDICTIONS

Additional municipalities in **Vermont** (Burlington), **Missouri** (St. Louis), **Pennsylvania** (Philadelphia), and elsewhere have passed resolutions of varying force. The total sits at roughly thirty and has been climbing steadily since 2019.

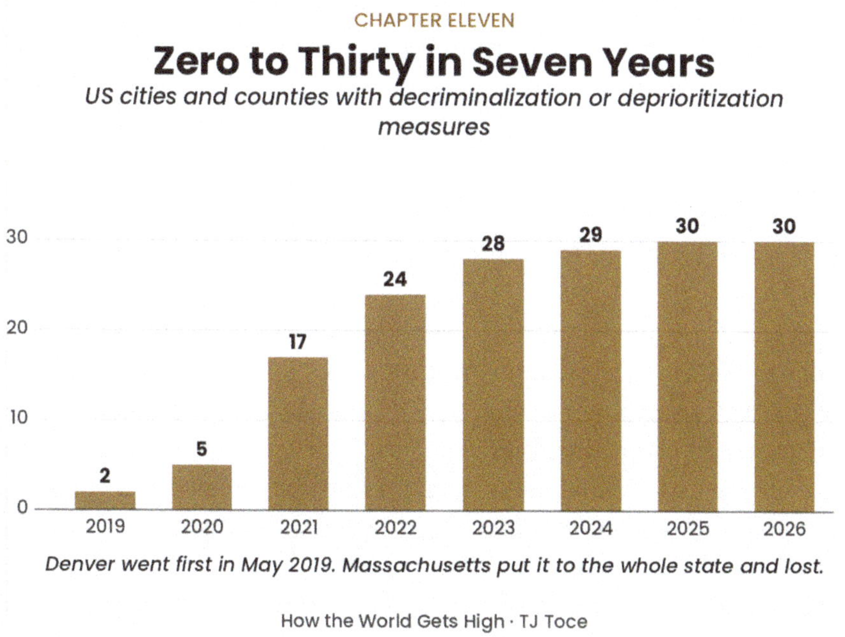

## PART III: WHAT IS MOVING NOW

Legislation worth watching as this book goes to press. Bills die, get amended and come back, so treat every line below as a snapshot rather than a status.

**1.** Multiple states have active bills to create regulated access or medical programs, with **Washington, Illinois, New York, Arizona, Missouri, Utah, Connecticut and Vermont** among the most frequently active.

**2. Texas** approved substantial state funding for ibogaine clinical development in 2025 and has continued to be an unexpectedly significant player in this field.

**3. Arizona** has funded psilocybin research through state appropriations.

**4.** Several states have created **psychedelic study commissions**, which is generally the step that precedes real legislation by two to four years.

**5.** Federal action beyond the FDA approval pathway remains unlikely in the near term. The realistic federal path is drug approval, not rescheduling by legislation.

---

### AN HONEST WORD ABOUT THIS CHAPTER

This is a point in time. Every list in this book will age. This one will age fastest.

At the pace of the last three years, some of what you just read will be out of date within months of publication. Cities will be added. A statewide program may launch. Something may be repealed, as Massachusetts voters reminded everyone in 2024.

So please treat this chapter as a photograph rather than a map. It shows you the shape of the thing and the direction it is traveling. It cannot tell you what is true in your town today.

**Before you rely on any of it, verify current local and state law.** Municipal resolutions can be rescinded by the next council. Prosecutorial policies change with the prosecutor.

---

## WHAT THE MAP ACTUALLY SHOWS

Step back and focus on the big picture.

In May 2019, the number of places in the United States where any government at any level had softened its position on psychedelics was **zero**.

Just seven years later: three states with regulated programs, the nation's capital, and roughly thirty cities and counties across a dozen states, red and blue, coastal and inland, in a country that agrees on almost nothing.

Denver passed by fewer than two thousand votes. D.C. passed with seventy-six percent of the vote.

That is not a fringe movement anymore. This is a shift in what the ordinary American thinks is reasonable, and it happened in under a decade.

And look at what is underneath it. Not one of these measures created a retail store or a recreational market. Almost every one of them was framed around healing. Around veterans. Around people who had tried everything else. Around the decent and logical argument to not put someone in jail for growing a mushroom that helped them stop wanting to die.

That argument turned out to be extremely persuasive.

Seven years. Zero to thirty-plus. The map is being redrawn day by day, one city council vote at a time.

Where will your town be on the next printing of this book? Use your voice and make them listen. You have more power than you realize.

Ciao for now!

# APPENDIX

# PART ONE

## THE DICTIONARY

### Every major psychedelic, and every other name it goes by

Here is a problem that trips up almost everyone.

A friend mentions huachuma. An article discusses mescaline. A retreat advertises San Pedro. A news story covers Colorado's mescaline provisions.

Those are all the same thing.

Somebody talks about the toad. Somebody else says 5-MeO. A researcher says Bufo. A veteran says the God molecule. Same compound, four names, and a reader could easily believe they are four different substances.

This dictionary exists to fix that. Each entry gives the scientific name, the source, and every other name you are likely to encounter, including street names, traditional names, clinical names and the names used in ceremony.

**A note on the street names.** They are included for recognition, not endorsement. Knowing that "molly" and "MDMA" are the same word is genuinely useful information, particularly for parents, teachers and clinicians. Many street names are also unreliable, because what is sold under a name is frequently not what the name means. Facts first, always.

### PSILOCYBIN

**Scientific:** Psilocybin (4-phosphoryloxy-N,N-dimethyltryptamine). Converts in the body to **psilocin**, which is the active compound. **Source:** Roughly 200 species of mushroom, principally the genus *Psilocybe*. **Chapter:** Two.

**ALSO KNOWN AS:** Magic mushrooms · Mushrooms · Shrooms · 'Shrooms · Boomers · Caps · Golden caps · Golden teachers (a specific *P. cubensis* variety, often used loosely) · Liberty caps (specifically *P. semilanceata*) · Blue meanies (specifically *P. cyanescens* and others) · Cubes (*P. cubensis*) · Philosopher's stones (specifically the sclerotia/truffles) · Truffles · Zoomers · Purple passion · Little smoke · Sacred mushrooms · Flesh of the gods · **Teonanácatl** (Nahuatl) · **Los niños santos**, the holy children (Mazatec, via Spanish) · **Ndi xijtho** (Mazatec) **Clinical/trade:** COMP360 (Compass Pathways synthetic formulation) · CYB003 · Psilocybin analog

## LSD

**Scientific:** Lysergic acid diethylamide. Also LSD-25. **Source:** Semi-synthetic, derived from ergot alkaloids (*Claviceps purpurea*). **Chapter:** Eight.

**ALSO KNOWN AS:** Acid · Lucy · Lucy in the sky · L · Tabs · Blotter · Blotter acid · Doses · Hits · Trips · Dots · Microdots · Sugar cubes · Windowpane · Sunshine · California sunshine · Orange sunshine · Yellow sunshine · Purple haze · White lightning · Cid · Sid · Stamps · Papers · Zen · Electric Kool-Aid **Trade:** Delysid (Sandoz, 1947 to mid-1960s)

⚠ **IMPORTANT:** Blotter paper sold as LSD is frequently something else, most dangerously the NBOMe family. See Chapter Eight.

## MDMA

**Scientific:** 3,4-methylenedioxymethamphetamine. **Source:** Entirely synthetic. **Chapter:** Eight.

**ALSO KNOWN AS:** Ecstasy · E · X · XTC · Molly · Mandy (UK) · Beans · Rolls · Rolling · Adam (the original therapeutic name, 1970s and 80s) · Disco biscuits · Love drug · Hug drug · Vitamin E · Scooby snacks · Thizz · Pills · Doves **Clinical:** Midomafetamine (the official generic name used in FDA filings)

⚠ **IMPORTANT:** "Ecstasy" pills and "molly" powder frequently contain other substances entirely, including methamphetamine, cathinones and synthetic opioids. The name tells you nothing about the contents.

**RELATED: MDA** (the "love drug," sassafras, Sally) is a distinct and more stimulating compound. **Methylone** (bk-MDMA, M1) is the compound Transcend Therapeutics is developing for PTSD.

## MESCALINE

**Scientific:** 3,4,5-trimethoxyphenethylamine. **Source:** Peyote (*Lophophora williamsii*), San Pedro (*Trichocereus pachanoi*), Peruvian torch (*Trichocereus peruvianus*), Bolivian torch (*Trichocereus bridgesii*). **Chapter:** Three.

**PEYOTE ALSO KNOWN AS:** Buttons · Peyote buttons · Cactus · Medicine · **Hikuri** (Wixárika/Huichol) · **Peyotl** (Nahuatl, the origin of the word) · Big chief · Mesc · Moon · Half moon

**SAN PEDRO ALSO KNOWN AS: Huachuma** · **Wachuma** · **Achuma** · **Aguacolla** · **Gigantón** · Saint Peter cactus · Cactus of the four winds · Trich · Pachanoi

⚠ **NOTE:** Peyote and San Pedro contain the same primary compound but are entirely different plants with entirely different legal, conservation and ethical situations. See Chapter Three.

### DMT

**Scientific:** N,N-dimethyltryptamine. **Source:** Hundreds of plant species, most notably *Psychotria viridis* and *Mimosa tenuiflora*. Also produced endogenously in mammals. **Chapters:** Three and Seven.

**ALSO KNOWN AS:** The spirit molecule · Dimitri · Businessman's trip · Businessman's lunch · Fantasia · The 45-minute psychosis (an old and unhelpful clinical term) · Changa (a smokable blend of DMT with an MAOI-containing herb) · Deems

### AYAHUASCA

**Scientific:** A brew of *Banisteriopsis caapi* (containing harmine, harmaline, tetrahydroharmine) plus a DMT-containing plant, usually *Psychotria viridis*. **Chapter:** Three.

**ALSO KNOWN AS: Yagé** / **Yajé** (Colombia, Ecuador) · **Caapi** · **Natem** (Shuar) · **Hoasca** (União do Vegetal) · **Daime** / **Santo Daime** (the Santo Daime church) · **Vegetal** (UDV) · **Nixi pae** (Huni Kuin) · **Uni** · **Kamarampi** (Asháninka) · La purga · The vine · Vine of the soul · Vine of the dead · The medicine · Grandmother · Mother Ayahuasca · Aya

**THE COMPONENT PLANTS:** *B. caapi* — the vine, ayahuasca proper, caapi, jagube, mariri *P. viridis* — **chacruna**, chacrona, the leaf, rainha (queen) *Diplopterys cabrerana* — **chaliponga**, chagropanga, oco yagé (an alternative DMT source)

### 5-MeO-DMT

**Scientific:** 5-methoxy-N,N-dimethyltryptamine. **Source:** The Sonoran Desert toad (*Incilius alvarius*), *Anadenanthera* tree seeds, various grasses. Also synthesized. **Chapter:** Four.

**ALSO KNOWN AS:** The toad · Toad medicine · **Bufo** (from the former genus name, extremely common though taxonomically outdated) · Bufo alvarius · Five-methoxy · 5-MeO · The God molecule · The Jaguar · Sapo (Spanish for toad, though see the warning below) · Yopo / Cohoba (referring to the *Anadenanthera* snuff, which also contains bufotenine)

⚠ **CRITICAL CONFUSION: "Sapo" is used for two completely different things.** In some contexts it means 5-MeO-DMT from the toad. In Amazonian contexts, *sapo* frequently means **kambo**, the frog secretion, which is not psychedelic at all. Clarify which is meant. See Chapter Four.

### KAMBO

**Scientific:** Peptide secretion of *Phyllomedusa bicolor*. **Chapter:** Four.

**ALSO KNOWN AS: Sapo** · **Vacina do sapo** (frog vaccine) · **Acate** (Matsés) · **Dow-kiet** · Frog medicine · Giant monkey frog · Waxy monkey tree frog

⚠ **NOT A PSYCHEDELIC.** No hallucinogenic effects whatsoever. Included here specifically because it is so persistently and wrongly grouped with them.

### SALVIA

**Scientific:** Salvinorin A, from *Salvia divinorum*. **Chapter:** Three.

**ALSO KNOWN AS:** Diviner's sage · Sage of the seers · Seer's sage · Magic mint · Sally-D · Sally · **Ska maría pastora** (Mazatec, "leaves of Mary the Shepherdess") · **Ska pastora** · **Yerba de María** · Hierba de la Virgen · Purple sticky (a marketing name for concentrated extracts)

⚠ **NOTE:** The only major psychedelic that does not act on the serotonin system. Entirely unlike the others in character.

### IBOGAINE

**Scientific:** Ibogaine, from *Tabernanthe iboga* root bark. **Chapter:** Three.

**ALSO KNOWN AS:** Iboga · **Eboga** · **Eboka** (Fang, Gabon) · Root bark · TA (total alkaloid extract) · Sacred wood · Bwiti medicine · The bitter root

**RELATED: Noribogaine** is the active metabolite. **Tabernanthalog** is the engineered non-hallucinogenic analog discussed in Chapter Eight.

## AMANITA MUSCARIA

**Scientific:** Muscimol and ibotenic acid, from *Amanita muscaria*. **Chapter:** Two.

**ALSO KNOWN AS:** Fly agaric · Fly amanita · The red mushroom · Soma (contested, see Chapter One) · **Mukhomor** (Russian) · **Wapaq** (Koryak) · Christmas mushroom · Santa's mushroom · Beni-tengu-take (Japanese)

⚠ **NOT PSILOCYBIN.** A completely different compound acting on a completely different receptor system. Legal in 49 US states.

## KETAMINE

**Scientific:** Ketamine hydrochloride. **Esketamine** is the S-enantiomer. **Chapter:** Eight.

**ALSO KNOWN AS:** K · Special K · Ket · Vitamin K · Kit Kat · Cat valium · Cat tranquilizer · Horse tranquilizer · Jet · Super acid · Green · Purple · Donkey dust **Trade:** Ketalar · **Spravato** (esketamine nasal spray) · Ketaset

⚠ **NOTE:** The only substance in this book legally prescribable in the United States for mental health treatment.

## LSA / MORNING GLORY

**Scientific:** Lysergic acid amide, also called ergine. **Source:** *Turbina corymbosa*, *Ipomoea tricolor*, *Argyreia nervosa*. **Chapter:** Three.

**ALSO KNOWN AS: Ololiuqui / Ololiuhqui** (Nahuatl) · **Badoh negro** (Zapotec) · Morning glory seeds · Heavenly Blue · Pearly Gates · Flying Saucers (all cultivar names) · Hawaiian baby woodrose · HBWR · Baby woodrose · Elephant creeper · Woodrose

## MAD HONEY

**Scientific:** Grayanotoxins, from *Rhododendron ponticum* and *R. luteum* nectar. **Chapter:** Six.

**ALSO KNOWN AS: Deli bal** (Turkish, "mad honey") · Grayanotoxin honey · Red honey · Cliff honey · Himalayan mad honey · **Pagal madhu** · Rhododendron honey · Poisonous honey · Xenophon's honey

### THE 2C FAMILY

**Scientific:** A family of phenethylamines, mostly designed by Alexander Shulgin. **Chapter:** Eight.

**MEMBERS AND NAMES: 2C-B** — Nexus · Bromo · Venus · Erox · Bees · Tucibi / Tusi (Latin America, though "tusi" is now frequently sold as a pink powder mixture containing anything at all) **2C-I** — Smiles (note: this name is also applied to the far more dangerous 25I-NBOMe) **2C-E** — Europa · Aquarust **2C-T-7** — Blue mystic · Lucky 7 · Tripstasy

⚠ **CRITICAL:** The name "smiles" has been applied to both 2C-I and 25I-NBOMe, which have dramatically different safety profiles. Deaths have resulted from this confusion.

### DATURA AND BRUGMANSIA

**Scientific:** Scopolamine, atropine, hyoscyamine, from *Datura* and *Brugmansia* species. **Chapter:** Three.

**ALSO KNOWN AS:** Jimsonweed · Jamestown weed · Devil's trumpet · Devil's snare · Thornapple · Moonflower · Angel's trumpet (*Brugmansia*) · **Toloache** (Nahuatl) · Locoweed · **Borrachero** (Colombia) · **Burundanga** / Devil's breath (referring to scopolamine)

⚠ **NOT A PSYCHEDELIC. A DELIRIANT.** Genuinely dangerous. See Chapter Three.

### SYRIAN RUE

**Scientific:** Harmine and harmaline, from *Peganum harmala*. **Chapter:** Three.

**ALSO KNOWN AS: Esfand / Espand** (Persian) · **Harmal** · **Üzerlik** (Turkish) · Wild rue · African rue · Mountain rue · **Hoama / Haoma** (contested link to soma) · Anahuasca ingredient · Pharmahuasca ingredient

### NUTMEG

**Scientific:** Myristicin, elemicin and safrole, from *Myristica fragrans.* **Chapter:** Three.

**ALSO KNOWN AS: Jaifal** (Hindi) · **Pala** (Indonesian) · Mace (the aril of the same seed, chemically similar) · Prison spice · Poor man's high · Nutmeg high

# PART TWO

## WHERE TO LEARN MORE

Every organization below is legitimate, publicly operating, and reachable. Inclusion is not endorsement, and nothing here is a recommendation to use anything. It is a map, nothing more.

### 1. RESEARCH INSTITUTIONS

**Johns Hopkins Center for Psychedelic and Consciousness Research** (Baltimore, MD) The most influential academic psychedelic research center in the world. Extensive public education materials. *hopkinsmedicine.org/psychiatry/research/psychedelics-research*

**NYU Langone Center for Psychedelic Medicine** (New York, NY) Psilocybin research for alcohol use disorder and cancer-related distress. *nyulangone.org*

**Imperial College London Centre for Psychedelic Research** (London, UK) The first formal academic psychedelic research center in the world, founded 2019. *imperial.ac.uk/psychedelic-research-centre*

**MAPS** — Multidisciplinary Association for Psychedelic Studies (San Jose, CA) Founded 1986. The organization that drove MDMA therapy through clinical trials. Extensive free archive and an annual conference. *maps.org*

**Usona Institute** (Madison, WI) Nonprofit medical research organization holding a National Priority Voucher for psilocybin in major depressive disorder. *usonainstitute.org*

**Heffter Research Institute** (Santa Fe, NM) Named for Arthur Heffter, who isolated mescaline in 1897. Funds psilocybin research worldwide. *heffter.org*

**Beckley Foundation** (Oxford, UK) Research and drug policy reform, founded by Amanda Feilding. *beckleyfoundation.org*

**Mount Sinai Center for Psychedelic Psychotherapy and Trauma Research** (New York, NY) *mountsinai.org*

## 2. TRAINING AND EDUCATION

**California Institute of Integral Studies** (San Francisco, CA) The Certificate in Psychedelic-Assisted Therapies and Research, the longest-running program of its kind. *ciis.edu*

**Naropa University** (Boulder, CO) Psychedelic-Assisted Therapies Certificate, contemplative tradition. *naropa.edu*

**Integrative Psychiatry Institute** (Boulder, CO) Psychedelic-assisted therapy training for licensed clinicians. *psychiatryinstitute.com*

**Fluence** (New York, NY) Continuing education in psychedelic integration for licensed therapists. *fluencetraining.com*

**Oregon Health Authority — Psilocybin Services** The state licensing authority. The place to start for facilitator training in Oregon. *oregon.gov/oha*

**Colorado Department of Regulatory Agencies — Natural Medicine Division** Colorado's licensing authority for facilitators and healing centers. *dora.colorado.gov*

## 3. CHURCHES AND RELIGIOUS ORGANIZATIONS

**Native American Church** The largest indigenous religious organization in North America. Federally protected peyote use for enrolled members of federally recognized tribes under the 1994 AIRFA amendments. Organized through regional and tribal chapters rather than a single national office.

**União do Vegetal (UDV)** Brazilian church using ayahuasca (hoasca) sacramentally. Won a unanimous US Supreme Court decision in 2006 establishing its federal religious exemption. US congregations in several states. *udvusa.org*

**Santo Daime** Brazilian syncretic church using ayahuasca (Daime). US churches in Oregon and elsewhere operate under federal court settlements. *santodaime.org*

**Zide Door Church of Entheogenic Plants** (Oakland, CA) Independent church founded 2019. Its legal position has been contested and its history includes law enforcement action. Included for completeness.

### 4. HARM REDUCTION AND SUPPORT

**Fireside Project** — Psychedelic Peer Support Line **Call or text 62-FIRESIDE (623-473-7433), daily, 11am to 11pm Pacific.** Free, confidential peer support during and after psychedelic experiences, and for anyone sitting with someone who is having one. This is the single most useful entry in this entire appendix. *firesideproject.org*

**DanceSafe** Drug checking, reagent test kits, and evidence-based education. Present at festivals and events nationally. *dancesafe.org*

**Zendo Project** Psychedelic peer support at festivals and events, and training in supporting difficult experiences. *zendoproject.org*

**Erowid** The largest independent archive of information and firsthand experience reports on psychoactive substances, online since 1995. *erowid.org*

**Psychedelic Support** Directory of licensed therapists offering psychedelic integration. *psychedelic.support*

### 5. POLICY, ADVOCACY AND ETHICS

**Chacruna Institute for Psychedelic Plant Medicines** (San Francisco, CA) The leading voice on indigenous reciprocity, ethics and social justice in this field. Essential reading for anyone thinking seriously about the questions raised in Chapter Nine. *chacruna.net*

**Indigenous Medicine Conservation Fund** Funds indigenous-led conservation of sacred plant medicines and the traditions around them. *imcfund.org*

**Psychedelic Medicine Coalition** (Washington, DC) Federal policy advocacy. *psychedelicmedicinecoalition.org*

**Decriminalize Nature** The grassroots organization behind the majority of the municipal measures listed in Chapter Eleven. *decriminalizenature.org*

**Students for Sensible Drug Policy** *ssdp.org*

**Drug Policy Alliance** *drugpolicy.org*

### 6. FURTHER READING

**1. *How to Change Your Mind*** — Michael Pollan (2018). The book that reopened this conversation for general readers, and the reason this book exists. Start here.

**2. *The Doors of Perception*** — Aldous Huxley (1954). Sixty pages. Still the best writing on the subject.

**3. *LSD: My Problem Child*** — Albert Hofmann (1979). The chemist's own account, and unexpectedly moving.

**4. *Food of the Gods*** — Terence McKenna (1992). Brilliant, wildly speculative, and to be read with a critical eye. Not everything in it holds up.

**5. *DMT: The Spirit Molecule*** — Rick Strassman (2000). The clinical data alongside the speculation. Read Chapter Seven of this book first for context.

**6. *PiHKAL*** and **7. *TiHKAL*** — Alexander and Ann Shulgin (1991, 1997). Half memoir, half chemistry manual. Nothing else like them.

**8. *The Immortality Key*** — Brian Muraresku (2020). The Eleusis argument, made at length and with genuine scholarship. Contested.

**9. *Sacred Knowledge*** — William Richards (2015). A Johns Hopkins researcher on the mystical experience, from inside the clinic.

**10. *Psychedelic Medicine*** — Richard Louis Miller (2017). Interviews with the major figures in the field.

**11. *Plants of the Gods*** — Richard Evans Schultes and Albert Hofmann (1979). The illustrated ethnobotanical reference. The foundational text.

**12. *María Sabina: Selections*** — edited by Jerome Rothenberg (2003). Her chants and her own words. Read it, and remember what this cost her.

### A CLOSING NOTE

If you take one thing from this appendix, make it the Fireside Project number.

**62-FIRESIDE. 623-473-7433. Daily, 11am to 11pm Pacific.**

Free, confidential, staffed by trained volunteers, available to anyone having a hard time during or after a psychedelic experience, and to anyone sitting with someone who is.

Everything else in this book is history, botany and wonder.

That number is the one piece of practical help in it. Facts first, and care always.

# PART THREE

## FINDING A PRACTITIONER

### What all those new "psychedelic therapists" actually are

Something has happened in the last five years that nobody quite planned. Search for a psychedelic therapist near you and you will find dozens. Some of them are physicians. Some are licensed under a brand new state profession that did not exist in 2020. Some are unlicensed and operating illegally. And some are doing something extremely valuable that involves no psychedelics at all.

They are not the same thing. They are not close to the same thing. And the reader most likely to be harmed is the reader who does not know the difference.

So here is the difference.

**One thing this section will not do is name anybody.** Clinics open and close constantly, and a list of names printed in a book is wrong within a year. What follows is the map of the categories, how to check anyone in any of them, and the directories that are actually maintained by people whose job it is to keep them current.

## THE FIVE CATEGORIES

### 1. KETAMINE CLINICS

**The only one you can walk into tomorrow.**

Ketamine is legal, prescribable, and has been an FDA-approved anesthetic since 1970. Esketamine, the nasal spray, has been approved for treatment-resistant depression since 2019. Because generic ketamine has been an approved medicine for over fifty years, physicians can prescribe it off-label, and thousands do.

There are now hundreds of ketamine clinics operating across all fifty states, alongside telehealth providers offering at-home protocols.

**What it looks like:** a medical setting. A physician or nurse practitioner. An IV infusion or a nasal spray. Vital signs monitored. Usually a series of sessions over several weeks.

**What to know:** this is a real medical treatment with genuine evidence behind it, and it is also a real industry with genuine oversight questions. Quality varies enormously between a psychiatrist-led clinic with integration support and a storefront running infusions on a conveyor belt. Insurance sometimes covers esketamine. It rarely covers generic ketamine infusions.

**Cost:** typically $400 to $800 per infusion, often six sessions to start.

### 2. LICENSED FACILITATORS IN OREGON, COLORADO AND NEW MEXICO

**A profession that did not exist five years ago.**

These are people who have completed a state-approved training program, passed an examination and hold a state license to supervise a psilocybin session. Nothing like this existed anywhere in the modern world before Oregon's first service center opened in 2023.

Oregon now has 22 licensed service centers and over 100 licensed facilitators. Colorado's healing centers opened in 2025 and already number 76. New Mexico's medical program is beginning enrollment.

**What it looks like:** a preparation meeting, a supervised session on site lasting around six hours, and an integration session afterwards. In Oregon no diagnosis and no doctor are required. Colorado's model is similar and also permits personal use. New Mexico requires a qualifying condition and physician involvement.

**What to know:** you must physically travel to the state and consume on site. You cannot take anything home. The facilitator is licensed by the state, which means there is a real regulator, a real complaints process and a real license to lose. That matters more than it sounds.

**Cost:** commonly over $1,000 per session and rarely covered by insurance. This is the single biggest criticism of these programs and it is a fair one.

### 3. INTEGRATION THERAPISTS

**The category almost nobody has heard of, and probably the most useful one.**

An integration therapist is a licensed psychotherapist, counselor or psychologist who helps you make sense of an experience you had somewhere else. They do not administer anything. They do not supply anything. They are not present during the experience.

They are legal in all fifty states and in most countries, because what they are doing is simply therapy.

**What it looks like:** ordinary talk therapy, with a therapist who is not going to be shocked, dismissive or obliged to file a report because you mentioned mushrooms.

**What to know:** the honest case for this category is that the experience is rarely the hard part. What comes afterward is. People routinely describe profound sessions that changed nothing, because nothing was built on top of them. An integration therapist is the person who helps you build.

If you are going to spend money on one thing in this list, many practitioners in this field would tell you to spend it here.

**Cost:** ordinary therapy rates. Sometimes covered by insurance, because it is billed as therapy, which is what it is.

### 4. RETREAT CENTERS ABROAD

**Legal where they operate, and enormously variable.**

Jamaica never scheduled psilocybin. The Netherlands sells truffles openly. Costa Rica, Mexico, Peru, Brazil and Portugal all host retreat industries operating in various shades of legality.

**What it looks like:** several days to a week, group setting, usually residential, often with preparation and integration built in.

**What to know:** the range of quality here is wider than in any other category, from centers with physicians on staff and rigorous medical screening to operations with none of that. The absence of regulation is the point of these jurisdictions and it cuts both ways.

Medical screening is the thing to insist on. A center that does not ask about your medications, your heart and your family psychiatric history before taking your deposit is telling you something important about itself.

**Cost:** commonly $2,000 to $8,000 for a week, not counting flights.

### 5. UNDERGROUND GUIDES

**Illegal, unregulated, and where the documented harm concentrates.**

Underground practitioners have existed continuously since 1970 and there are far more of them than all the licensed categories combined. Some are experienced, careful and have been doing this responsibly for decades. Some are not, and there is no way to tell them apart from a website.

**What to know, plainly:**

There is no license, so there is nothing to revoke. There is no regulator, so there is nowhere to complain. There is no insurance and no recourse. And the reported cases of serious misconduct in this field — including sexual misconduct during sessions — come overwhelmingly from unregulated settings where the practitioner faced no professional consequence.

This section is not going to lecture you. But it would be dishonest to list this category alongside the others without saying that the protections which make the first three categories reasonably safe are precisely the protections this one lacks.

## HOW TO CHECK ANYONE, IN ANY CATEGORY

This is the part worth photographing and keeping.

### Ask these seven questions

**1. What is your license, and in what state?** A licensed clinician or facilitator will answer immediately and specifically. Then verify it yourself. Every US state has a free online license lookup for physicians, psychologists, counselors and social workers. Oregon and Colorado both publish searchable registries of licensed psilocybin facilitators. Takes two minutes.

**2. What is your medical screening process?** Anyone competent will screen for cardiac conditions, for personal and family history of psychosis or bipolar disorder, and for medication interactions — particularly SSRIs, MAO inhibitors and lithium. **If nobody asks you about your medications and your family psychiatric history, walk away.**

That single question separates the serious from the rest better than any other.

**3. What happens if something goes wrong?** There should be an actual answer. Who is present, what the medical plan is, how far the nearest hospital is, whether anyone on site is trained in managing an emergency.

**4. What does integration look like, and is it included?** Someone who treats the session as the whole product and integration as an upsell has the emphasis backwards.

**5. Can I speak to someone who has worked with you?** Reasonable request. A defensive reaction to it is informative.

**6. What is the total cost, in writing?** Including preparation, the session itself, integration, and anything else.

**7. What are your boundaries around physical contact?** This is an awkward question and you should ask it anyway. There should be a clear, unembarrassed, professional answer, given in advance and in writing.

### Six warning signs. Any one of them is enough.

**1. No written agreement.** No informed consent document, no terms, nothing you can read before you commit.

**2. No medical screening.** Covered above, and it is the most common failure.

**3. Discouraging you from telling anyone.** Legitimate practitioners understand you may want to talk to your doctor, your therapist or your family. Anyone who frames secrecy as part of the process is protecting themselves, not you.

**4. Any romantic or sexual dimension, at any point.** Not during, not before, not after, not "when the work is complete." This is the single most serious documented harm in this field. There is no framework in which it is appropriate, and there is no explanation worth listening to.

**5. Claims that cannot be true.** Guaranteed outcomes. Cures for named conditions. Any suggestion you should stop taking prescribed medication.

**6. Pressure.** Urgency, limited availability, discounts that expire, or discomfort when you say you want to think about it. Nobody doing this responsibly needs you to decide today.

## DIRECTORIES THAT STAY CURRENT

These are maintained by organizations whose job it is to keep them updated, which is why they are here instead of a list of names that would age badly.

**Oregon Health Authority — Psilocybin Services.** The official state registry of licensed facilitators and service centers. *oregon.gov/oha*

**Colorado Department of Regulatory Agencies — Natural Medicine Division.** The official state registry of licensed facilitators and healing centers. *dora.colorado.gov*

**Your state's professional licensing board.** Every state publishes free lookup tools for physicians, psychologists, counselors and social workers. Search your state's name plus "license verification."

**Psychedelic Support.** Directory of licensed therapists offering integration work. *psychedelic.support*

**MAPS.** Maintains information on clinical trials, which remain one legitimate route to access. *maps.org*

**ClinicalTrials.gov.** Search by condition and location. Enrolling in a trial means treatment under research conditions, usually at no cost.

**Fireside Project.** Free, confidential peer support by phone or text. **62-FIRESIDE, or 623-473-7433.** Not a directory, but the number to have if a session becomes difficult or if you are sitting with someone who is struggling.

## AND ONE LAST THING

The whole of this book has argued that humans have been doing this for ten thousand years, almost always in the presence of somebody who knew what they were doing. A Mazatec curandera singing through the night. A Bwiti initiation with the whole village present. A Gurung honey hunter who learned the cliff from his father.

The container was never optional. It was the technology.

What is new in 2026 is not the substances. It is that a container is being rebuilt in public, with licenses and regulators and complaints procedures, for the first time in the modern West.

Choose your company carefully. That part has never changed.

**Facts first. And your own good judgement always.**

# PART FOUR

## THE SHAMAN QUESTION

### Traditional, non-traditional, and how to tell which is which

Search for a shaman in America and you will find one within twenty miles of wherever you are sitting. There are shamans in Brooklyn and shamans in Boulder and shamans in Sedona and shamans, one assumes, in Des Moines. Some of them are the real article. Some of them are something else entirely, which is not automatically a bad thing. And some of them are even wearing a costume.

This is the most delicate territory in the book, and it deserves the same treatment everything else has had. Facts first.

## START WITH THE WORD

**"Shaman" is a Siberian word.**

It comes from *šaman* in Evenki, a Tungusic language of the Siberian taiga. Russian explorers heard it in the seventeenth century and carried it home. Anthropologists then did something strange with it: they took a specific job title from a specific people and applied it to roughly a thousand unrelated traditions across every continent on earth.

So it is worth knowing that almost nobody who does this work traditionally calls themselves a shaman.

A Mazatec healer in Oaxaca is a **curandera**, or *sabia*, a wise one. A Shipibo practitioner in the Peruvian Amazon is an **onanya**, one who knows, or a **meraya** at the highest level. A mestizo Amazonian healer is a **vegetalista** or an **ayahuasquero**. A Native American Church ceremonial leader is a **roadman**. A Bwiti initiator in Gabon is an **nganga**.

None of them are shamans. They are all doing something in a specific tradition with a specific name, learned over a specific number of years from a specific person.

**Which produces a rule of thumb that is uncomfortable but useful.** In the United States, the word "shaman" on a business card is closer to an inverse credential than a credential. The people with the deepest traditional training are the least likely to use it, because it is not their word. This is not a slur on anyone. It is just how the language shook out.

## TRADITIONAL

What is genuinely practiced in America, and what the law says about it

### The Native American Church

**The oldest, the largest, and the only one with real federal protection.**

The Native American Church is not new and not fringe. It has existed in organized form since 1918, has an estimated membership in the hundreds of thousands across dozens of tribal nations, and conducts all-night ceremonies built around peyote, prayer, water, fire, song and a drum.

The 1994 amendments to the American Indian Religious Freedom Act make peyote use for bona fide ceremonial purposes legal under federal law **for members of federally recognized tribes.**

**It is not open to everyone, and that is not a snub.** It is a specific religious practice belonging to specific communities who fought in court for the better part of a century to keep it. The reader who finds that disappointing is invited to notice how recently these ceremonies were prosecuted, and by whom.

### The peyote crisis, and the one thing this book will ask of you

Peyote is in trouble.

It is a slow plant. A button can take a decade or more to reach ceremonial maturity in the wild, and it grows in a narrow band of south Texas and northern Mexico and nowhere else on earth. It has been hit from every direction at once: root-plowing for agriculture, oil and gas development, loss of access to private ranch land, poaching, and harvesting by people who cut it wrong and kill the root.

The Native American Church declared this a crisis in 1995. In 2017 the **Indigenous Peyote Conservation Initiative** was formed to answer it, and now runs a 605-acre preserve near Hebbronville, Texas, doing seed banking, sustainable-harvest research and cultural education, on a timeline measured in fifty years.

Here is the part that matters. **Indigenous leaders have publicly and repeatedly asked the psychedelic movement to leave peyote alone** — including asking that peyote be specifically carved out of decriminalization bills, which is the opposite of what most reform advocates instinctively want.

The reasoning is simple arithmetic. There is not enough peyote for the people whose religion requires it. Adding several million curious newcomers to the demand for a plant that takes ten years to grow does not end well for anybody.

There is mescaline in San Pedro, which grows fast, grows tall, grows almost anywhere, and carries no such burden. Chapter Three covered both. **If mescaline is what you are after, the cactus you want is the Peruvian one.**

That is the only request this book makes of its reader, and it is not really a hard one.

### The ayahuasca churches

**Legally protected, and genuinely open.**

In 2006 the Supreme Court ruled unanimously in *Gonzales v. O Centro Espírita Beneficente União do Vegetal* that the UDV's ayahuasca sacrament was a sincere religious exercise protected by the Religious Freedom Restoration Act. In 2009 a federal judge in Oregon applied that precedent to the Santo Daime church. The Iowaska Church of Healing reached a settlement with the DEA in 2024, and in 2025 the Church of Gaia became the first to win an exemption through petition rather than litigation.

These are Brazilian syncretic Christian churches, hymn-based and highly structured, and they admit newcomers.

**They are, as of this writing, the only setting in which an ordinary American can legally take a classical psychedelic in a ceremonial context without traveling to Oregon or Colorado.** That is a genuinely useful fact and very few people know it.

### Visiting lineage practitioners

Shipibo, Mazatec, Kichwa and other Amazonian and Mexican healers travel a real circuit through American cities, invited by communities here, running ceremonies in private homes and rented retreat spaces.

Some of these are exactly what they appear to be: a person with thirty years of training doing the work they were raised in.

**Two honest cautions.** The legal protection described above does not extend to them, so both practitioner and participant are exposed. And the economics are brutal in a specific way — the sums an American ceremony can generate are transformative in the Amazon, which creates an enormous incentive for people with little or no training to present themselves as lineage healers. The genuine ones and the invented ones use the same words on the same kind of website.

## NON-TRADITIONAL

The Western shamanic movement, which is a real thing and not the same thing

### Core shamanism

In 1980 an American anthropologist named Michael Harner published *The Way of the Shaman* and later founded the Foundation for Shamanic Studies. His idea was deliberate and explicit: extract a "core" set of techniques common to shamanic traditions worldwide — drumming, the journey, spirit contact — and teach them stripped of any particular culture.

It worked, commercially and culturally. Most of the weekend workshops, drum circles, certification programs and practitioners you will encounter in America descend from this lineage or one like it.

**This book's position is that this deserves to be described accurately rather than sneered at.** Core shamanism is a Western spiritual practice, roughly fifty years old, with sincere practitioners and some very good ones, and people get real value from it. Notably, most of it involves no psychedelics whatsoever — the drum does the work.

The trouble is not that it exists. The trouble is when it is sold as something it is not.

### Plastic shamans

That is the term Native activists use, and they have been using it for a long time. In 1993 an assembly of Lakota, Dakota and Nakota representatives issued a formal **Declaration of War Against Exploiters of Lakota Spirituality** — an actual document, still worth reading, aimed squarely at non-Native people selling sweat lodges, vision quests and pipe ceremonies for money.

The strongest argument for taking that seriously is not an abstract one about appropriation. It is that people have died.

In October 2009, at a Sedona retreat run by a self-help entrepreneur with no traditional training, three people died in an overcrowded, overheated sweat lodge. He was convicted on three counts of negligent homicide. A real sweat lodge is run by someone who has spent years learning how not to kill anyone in one. That knowledge is the tradition. It is not decoration and it is not transferable in a weekend.

## HOW TO TELL THE DIFFERENCE

Four questions and one tell.

**1. Who trained you, by name, and for how long?** A real answer names a person, a place and a number of years. Vagueness here — "many teachers," "the ancestors," "I was called" — is the answer.

**2. What tradition is this, specifically?** Not "shamanic." Which people, which region, which lineage.

**3. Where does the medicine come from, and what did you pay for it?** Anyone serious knows the answer and will tell you.

**4. What is your relationship to the community this comes from?** Ongoing and specific, or a trip they took once.

**And the tell, which is worth more than the four questions combined: watch for mixing.**

A practitioner offering Peruvian ayahuasca, Mexican mushrooms, Amazonian rapé, a Native American sweat lodge and Tibetan sound healing in the same weekend has not trained in five traditions. Nobody has.

Each of those represents a decade of apprenticeship in a different language on a different continent. **A menu is not a lineage.** Real practitioners are almost always narrow, and slightly boring about it, and modest about what they do not do.

## TAKE THE HIGH ROAD

The word gets used loosely. In practice it means four things.

**Pay properly and do not haggle.** This is somebody's life's work.

**Ask where the medicine came from.** Wild-harvested, cultivated, purchased from whom.

**Do not take peyote out of Native hands.** Covered above, and it is the concrete one.

**Support the source communities directly.** The Indigenous Peyote Conservation Initiative, the Chacruna Institute's reciprocity work and ICEERS all do this, and all of them are easy to find.

## AND THE POINT OF ALL OF IT

This book has spent ten chapters with people who did this inside a tradition. A Mazatec curandera who sang for seven hours. A Bwiti initiation the whole village attended. A Gurung honey hunter who learned the cliff from his father, who learned it from his.

In none of those cases was the tradition an accessory to the experience. **The tradition was the safety equipment.** It was ten thousand years of accumulated knowledge about what to do when things go sideways at three in the morning, compressed into songs and rules and the steady presence of somebody who had seen it before.

You can absolutely find that in America. It is here, it is real, and some of it is even legal.

It just does not usually advertise.

**Facts first. And ask who taught them.**

# SELECTED SOURCES

This is not a full bibliography. It is a list of the principal works and studies behind each chapter, offered for readers who want to go further and as an account of where the claims in this book come from.

Where a specific figure, date or finding appears in the text, it is drawn from one of the sources below.

## GENERAL

Pollan, Michael. *How to Change Your Mind.* Penguin Press, 2018.

Schultes, Richard Evans, and Albert Hofmann. *Plants of the Gods: Their Sacred, Healing and Hallucinogenic Powers.* McGraw-Hill, 1979. The foundational ethnobotanical reference for this entire subject.

Erowid. Independent archive of firsthand reports and substance information, online since 1995. *erowid.org*

### CHAPTER ONE — THE OLDEST APPETITE

Samorini, Giorgio. Published analysis of the Tassili n'Ajjer rock art, *Integration*, 1992. The mushroom interpretation is contested and the text says so.

Akers, Brian P., et al. "A Prehistoric Mural in Spain Depicting Neurotropic *Psilocybe* Mushrooms?" *Economic Botany*, 2011. The Selva Pascuala argument.

Wasson, R. Gordon. "Seeking the Magic Mushroom." *Life*, 13 May 1957.

Wasson, R. Gordon, Albert Hofmann and Carl A. P. Ruck. *The Road to Eleusis.* Harcourt Brace Jovanovich, 1978. The ergot hypothesis for the kykeon, unproven.

Sahagún, Bernardino de. *Florentine Codex: General History of the Things of New Spain*, c. 1540–1585, on teonanácatl.

Miller, Melanie J., et al. "Chemical Evidence for the Use of Multiple Psychotropic Plants in a 1,000-Year-Old Ritual Bundle from South America." *PNAS*, 2019. The Cueva del Chileno bundle.

Terry, Martin, et al. Radiocarbon dating of the Shumla Caves peyote specimens, *Journal of Archaeological Science*, 2006.

Estrada, Álvaro. *María Sabina: Her Life and Chants.* Ross-Erikson, 1981.

## CHAPTER TWO — FUNGI

Hofmann, Albert. *LSD: My Problem Child.* McGraw-Hill, 1980. First-person account of the isolation of psilocybin and the synthesis of LSD.

Lenz, Claudius, et al. Research identifying the enzymes and oligomers responsible for the blue bruising reaction in *Psilocybe*, published in *Angewandte Chemie* and related journals, 2019–2020.

Reynolds, Hannah T., et al. "Horizontal gene cluster transfer increased hallucinogenic mushroom diversity." *Evolution Letters*, 2018.

Caporael, Linnda R. "Ergotism: The Satan Loosed in Salem?" *Science*, 1976. And the rebuttal: Spanos, Nicholas P., and Jack Gottlieb, *Science*, 1976. Both are cited because the dispute is the point.

Oregon Health Authority, Psilocybin Services programme documentation.

## CHAPTER THREE — PLANTS

McKenna, Dennis J. Published work on the pharmacology of ayahuasca and the MAO-inhibition mechanism.

Huxley, Aldous. *The Doors of Perception.* Chatto & Windus, 1954.

Lotsof, Howard. Accounts of the 1962 ibogaine observation, and subsequent published research on ibogaine in opioid dependence.

Cherian, Kirsten N., et al. Stanford study of ibogaine treatment in veterans with traumatic brain injury, *Nature Medicine*, 2024.

Terry, Martin, and Keeper Trout. Published work on peyote conservation and the decline of wild populations in south Texas.

Native American Church statements on peyote conservation and on requests to non-Native seekers, collected by the Indigenous Peyote Conservation Initiative.

American Indian Religious Freedom Act Amendments of 1994, Public Law 103-344.

## CHAPTER FOUR — AMPHIBIANS

Weil, Andrew T., and Wade Davis. "Bufo alvarius: a potent hallucinogen of animal origin." *Journal of Ethnopharmacology*, 1994.

Davis, Alan K., et al. Johns Hopkins survey research on 5-MeO-DMT and reported personal meaningfulness, *Journal of Psychopharmacology*, 2018–2019.

Most, Albert. *Bufo alvarius: The Psychedelic Toad of the Sonoran Desert*, 1984. The pamphlet that began the modern practice.

Arizona Game and Fish Department and New Mexico Department of Game and Fish, species status listings.

### CHAPTER FIVE — FISH

de Haro, Luc, and Philip Pommier. "Hallucinatory fish poisoning (ichthyoallyeinotoxism): two case reports from the western Mediterranean and literature review." *Clinical Toxicology*, 2006.

Pliny the Elder. *Naturalis Historia*, c. 77 CE.

### CHAPTER SIX — HONEY

Xenophon. *Anabasis*, Book IV, c. 370 BCE.

Strabo. *Geographica*, Book XII, on the Heptakometes and Pompey's soldiers.

Gunduz, Abdulkadir, et al. Extensive published clinical literature on grayanotoxin poisoning from the Karadeniz Technical University group, Trabzon, Türkiye.

Valli, Eric, and Diane Summers. *Honey Hunters of Nepal.* Harry N. Abrams, 1988.

### CHAPTER SEVEN — THE HUMAN BODY

Strassman, Rick. *DMT: The Spirit Molecule.* Park Street Press, 2000. The clinical data is solid; the pineal hypothesis is speculative and the text says so.

Dean, Jon G., et al. "Biosynthesis and Extracellular Concentrations of N,N-dimethyltryptamine (DMT) in Mammalian Brain." *Scientific Reports*, 2019.

Timmermann, Christopher, et al. "DMT models the near-death experience." *Frontiers in Psychology*, 2018.

Devane, William A., Lumír Hanuš, Raphael Mechoulam, et al. Isolation and naming of anandamide, *Science*, 1992.

Grof, Stanislav. *Holotropic Breathwork.* SUNY Press, 2010.

### CHAPTER EIGHT — SYNTHETIC

Hofmann, Albert. *LSD: My Problem Child.* McGraw-Hill, 1980.

Shulgin, Alexander and Ann. *PiHKAL: A Chemical Love Story*, 1991, and *TiHKAL: The Continuation*, 1997. Transform Press.

Berman, Robert M., et al. "Antidepressant effects of ketamine in depressed patients." *Biological Psychiatry*, 2000.

Cameron, Lindsay P., et al. "A non-hallucinogenic psychedelic analogue with therapeutic potential." *Nature*, 2021. Tabernanthalog.

United States Senate Select Committee (Church Committee) reports on MK-Ultra, 1975–1976.

### CHAPTER NINE — 2026 AND BEYOND

Griffiths, Roland R., et al. "Psilocybin can occasion mystical-type experiences having substantial and sustained personal meaning and spiritual significance." *Psychopharmacology*, 2006. The study that reopened the field.

Bogenschutz, Michael P., et al. "Percentage of Heavy Drinking Days Following Psilocybin-Assisted Psychotherapy vs Placebo in the Treatment of Adult Patients With Alcohol Use Disorder." *JAMA Psychiatry*, 2022.

Johnson, Matthew W., et al. Johns Hopkins psilocybin smoking cessation studies, *Journal of Psychopharmacology*, 2014, and follow-up work.

Krebs, Teri S., and Pål-Ørjan Johansen. "Psychedelics and Mental Health: A Population Study." *PLOS ONE*, 2013. 130,152 respondents, no association found.

Mitchell, Jennifer M., et al. MAPS phase 3 MDMA-assisted therapy trials, *Nature Medicine*, 2021 and 2023.

US Food and Drug Administration. Guidance for industry on psychedelic drug clinical investigations, final version, 2026.

CONFEN (Brazil). Working group report on ayahuasca religious use, 1987.

### CHAPTER TEN — A HAWAIIAN INTERLUDE

Handy, E. S. Craighill, and Elizabeth Green Handy. *Native Planters in Old Hawaii*. Bishop Museum Press, 1972.

Johnston, Ed, and Helen Rogers. *Hawaiian 'Awa: Views of an Ethnobotanical Treasure*. Association for Hawaiian 'Awa, 2006.

Sarris, Jerome, et al. Published pharmacological reviews of kavalactone mechanisms of action.

Hawai'i Senate Bill 3199 (2026).

Twain, Mark. Speech at the Delmonico's banquet honoring the Spalding world baseball tour, New York, April 8, 1889. Reported in the *Boston Daily Globe*, April 9, 1889, p. 1; printed in Harry Clay Palmer, *Athletic Sports in America* (Philadelphia: Hubbard Brothers, 1889), 444–47; collected in *Mark Twain's Speeches* (New York: Harper & Brothers, 1923).

*Legal status, trial phases and municipal counts reflect the position as of publication and will change. Verify anything you intend to rely on.*

## CHAPTER ELEVEN — THE MAP REDRAWN

Oregon Ballot Measure 109 (2020); Colorado Proposition 122 (2022) and Senate Bill 23-290 (2023); New Mexico Senate Bill 219 (2025); District of Columbia Initiative 81 (2020).

Municipal ordinances and resolutions as enacted in the cities listed. Decriminalize Nature maintains a running record.

# ABOUT THE AUTHOR

TJ Toce is a career retail executive who has always been curious about psychedelics and could not stop pulling the thread. *How the World Gets High* is the result: a global tour of sacred mushrooms, Amazon vines, mad honey and desert toads, and the science that is finally catching up to nine thousand years of human practice.

He is not a scientist. He is a very persistent seeker on that millennia-old path to enlightenment.

TJ lives in Beverly Hills, where he is at work on his next book.

*Keep up with what changes next at*
**howtheworldgetshigh.com**
*a running record of psychedelic law and research, updated as it happens.*